The Rain Maker Device

The Rain Maker Device
A Path to Harness Mother Nature

by
**Kosol and Koeun Ouch,
Vince Panella
and
David Lowrance**

E-BookTime LLC
Montgomery Alabama

The Rain Maker Device
A Path to Harness Mother Nature

Copyright © 2006 by Kosol and Koeun Ouch, Vince Panella and David Lowrance

All rights reserved. No part of this book may be reproduced or transmitted in any form or by any means, electronic or mechanical, including photocopying, recording, or by any information storage and retrieval system, without permission in writing from the copyright owner.

ISBN: 1-59824-374-8

First Edition
Published October 2006
E-BookTime, LLC
6598 Pumpkin Road
Montgomery, AL 36108
www.e-booktime.com

Dedication

This book is dedicated with special thanks to:

Carlos Sanchez (Best friend, long time student and colleague.)

Jessy Sanchez (Long time friend and student.)

Koeun Noun Ouch, David Lowrance, Jake Tepac, Tim, Jerry Evans II, Ben, Vince Penala (Italy), Daniel Nissen, Deltas, Martin and Lacosta (Spain).

Without their help and support this book would not have been possible.

Contents

Introduction and the Conscious Energy Loop 9
Phi, Numerology, and Density 14
Zero Point and the Five Sources of Magnetism 26
A Model of the Human Life Form 30
Time Vector Model of Magnetism and the Aether 39
Working With Conscious Energy 55
Vortex Analysis 59
Measuring the Background Aether Pressure,
    Density Factor 78
Protonics 81
Force Vector Model of the Atom 100
Body of Experimental Evidence Offered by
    C_S_S_P Group 112
Magnetism 232
Force Vector Model of the Atom 297
Creation of the Physical Fabric 306
More on Creation of the Physical Fabric 322
Magnetic Vortex Generator - Construction Detail 342
Conclusion 349
Diamagnetic Fields Without Generators 369
David Lowrance Testimony 372
Final Thoughts 386
Photographs 389

Introduction and the Conscious Energy Loop

On Entrophy and Awareness

"Entropy is a state of negative mind, an illusion." All is motion, and motion can not be stopped at the deepest levels of matter." This is a true Law of nature. "Nothing comes to rest, ever". "All is change, and change is constant." There is no such thing as an "isolated system," all systems composed of atomic particles are connected to forces we can not at present understand, all atomic particles move through transitions of disappearing and reappearing in this realm. Consciousness must slide deeper to realize these truths. Awareness is not fixed so as to be able to determine exact truth for all. A scientific model needs to accept this parameter. This is open ended science, one that accepts its present understanding is not a "law" but is fluid, just as reality is fluid to the level of awareness achieved." "Experiments involving Awareness or Consciousness do not produce consistent results for every one doing them, yet they should not be dismissed for this reason alone, but awareness must itself become a variable of the experiment."

When I sat up this site it was my goal to delve deeply into Magnetism to find the connection to Source power that has been hinted at from the past.

From the start, and not believing that energy can magically come from no where, I sat out to locate [Source] the basic energy that powers our universe, which manifests an inexhaustible supply feeding all atoms in our realm. I have labeled it as merely Source in all my personal work.

This Source of regenerative energy is presently called ZPE or Zero Point Energy in the scientific community. I never dreamed I would discover a form of Conscious Energy, the CU or consciousness unit, how to create one, or the connections to the inner planes of awareness referred to as the control fabric and the perception fabric introduced by Wilbert Smith.

I have tried to leave a solid record of the process I have followed, the perceptions gained, and the methods I used to reach personal comprehension.

If the world of mankind learns to use this Conscious Energy, to learn directly from it, it will change mankind forever.

The basics are found to be very simple. The inflow and the outflow can be generated using the Proton and Electron magnetic fields. Either an inflow or an outflow can be set up in a symmetrical fashion using a magnetic field looping patterned below in the diagram. The conscious energy loop set up in this way will continue to function even after the device is shut down and removed. It will continue to be reactive to the "intention". Outflow can be drawn off the free standing CU after it is established and interacted with as a scalar interaction. Depending on the direction of flow when it is activated, either higher time flow rate or lower time flow rate is set up.

After the third day following the removal of the device in experiment 7, "Rain Maker 1" I have no more doubts as to the reality of the energy field as being permanently established within the space it was created in.

I have now charted the flows appearing in the free standing self powered field. They follow the devices imprint exactly and now entering their third day of self sufficient flow in free space with no device to aid them. The scalar field is warm to the hands and has a spherical shape around the system.

The Vortex Flow Pattern

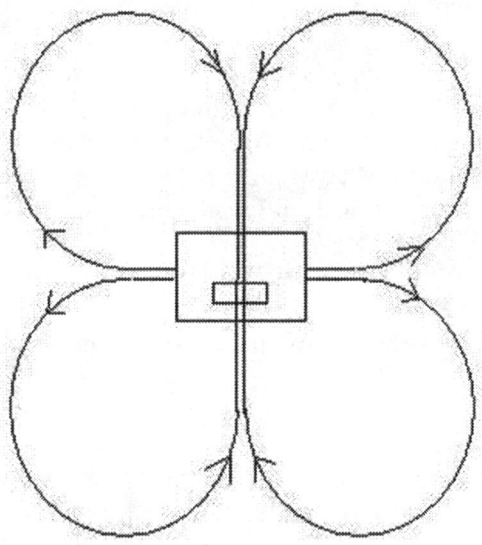

Much remains to be understood about this form of conscious energy, however I have found first hand that the inflow side, the perception fabric, has its own form of teaching you personally.

The book begins with a collection and compression of some of my personal theories of magnetism, and then moves to the experimental level, then finally to the discovery of the inflow and outflow energy system still active here in my upstairs work room.

The Conscious energy loop, just like the vortexes at Sedona Arizona, do not deflect the needle of a compass in the slightest and yet they are easily detected by any experienced meditator. They would appear to be EM [electro magnetic] scalar canceling or [Protonic fields] by nature, and this is why the human can access them and the compass cannot.

A Short Summary

The basic principles:

Gravity is the result of an altered time flow rate in matter.

The Proton sits in the strong force area of the atom in an altered time frame where gravity is slightly lower, it bridges the strong force and the EM layers.

The Electron sits in a different time frame where gravity matches here.

Torsion or "time flow rate" is coupled through a magnetic field.

The materials Copper Aluminum and Bismuth are magnetic at the Proton layer.

The materials Iron and Nickel are magnetic at the Electron layer.

Lightspeed is higher through the Protonic propagation medium [Aether] than the normal "free space" Electron medium [Either].

Tesla suggested the Time flow rate differential is around 1.5708 between Proton and Electron propagation velocities from direct experimental observation. This puts Protons field fabric in an alternate density and explains why Gravity does not interact with a normal electron magnetic field to form an interference pattern.

Coupling a Proton magnetic material to an Electron magnetic material with a magnetic field allows the two time frames to interact as a function of magnetic field strength and time through the B field which spans both "mediums". Using a scalar cancelling coil to reduce the Electrons magnetic field it becomes a mirror reflecting the Protons magnetic field back inwards.

If this process is followed into a spherical or cylindrical form, a spherical field is set up altering the time flow rate of the sphere.

The Rain Maker Device

If we alter it the correct direction then gravity should be lowered.

The magnets polarity determines whether the field is a slower time frame or a higher one as it couples the Electron to the Proton in only two possible methods.

The result is a spherical altered time flow rate, the CU or Consciousness Unit with properties that seem to also respond directly to intention in the Conscious fabric. It forms a time flow sheer face along its corona just as the strong force area of the atom, dropping off almost instantly at its edge. The actual CU is not magnetic but "scalar cancelling" or "Protonic" in a sphere of altered time flow rate.

The study has led to a new model of Protonics being developed through direct experiment, which is attempting to define the parameters of time flow rate and gravity fields. Also redefining the source of the diamagnetic field to be centered not in the Electron shell but the Protons.

I, and all working in the c_s_s_p group and the Kosol_Core_Tec group, hope you find this book interesting.

Phi, Numerology, and Density

Phi

Phi represents the relationship between "density" time vectors, and is the path inwards to higher density in physical reality.

This is related to the stability that matter takes in each of its forms or densities, along a "spiral" as energy moves in and out of Source, through all the Densities.

Time flow rate progresses along the spiral, increasing in its linear dimensions and slowing, related by Phi and 1/Phi. [Intuitive principle, consciousness or connection to Source as we move inwards.]

This idea is based on Kosol's insights as to the "math" of higher Consciousness [Phi].

Consider the basic phi spiral, a pattern of moving inwards towards Source.

The path by which all energy moves to and from Source, through all the Densities of reality.

If we examine the radius of the arc found in each box below we discover the importance of $1/Phi$, $1/Phi^2$, $1/Phi^3$...etc.

Spheres of force forming in each density would have various radius based upon the progressive increasing of the radius, the linear tempic vector, as it increases outwards. The relationship of radius increasing by Phi can be used to represent the tempic vector creating all matter along any Density.

How can this relationship represent the connection between forces interacting across densities?

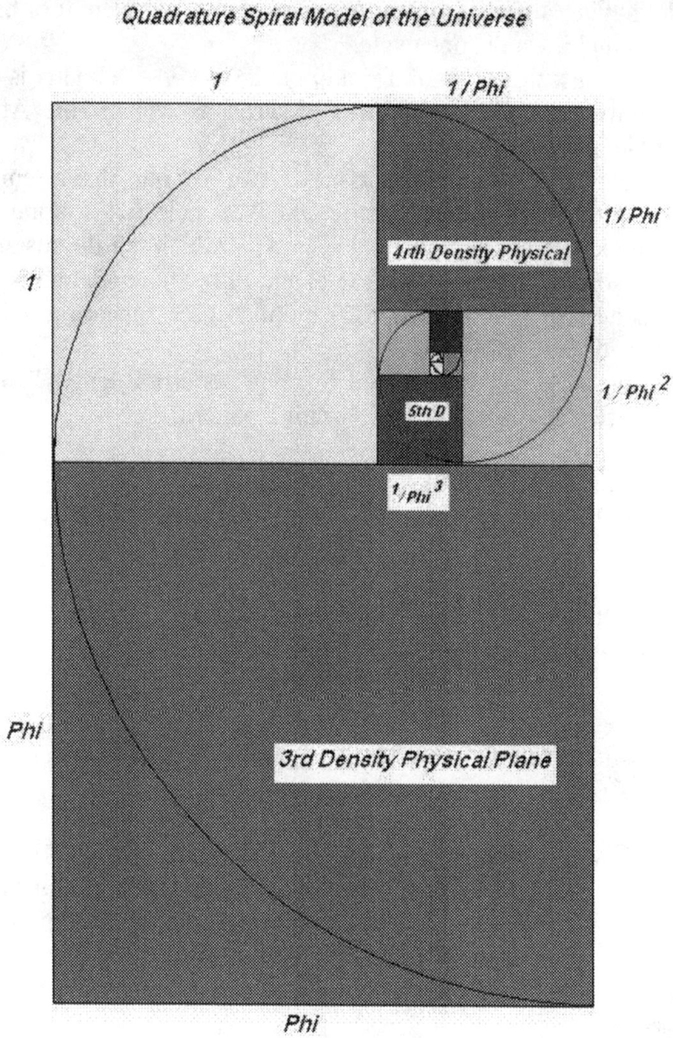
Quadrature Spiral Model of the Universe
1
1 / Phi
1 / Phi
4rth Density Physical
1
1 / Phi 2
5th D
1/Phi 3
Phi
3rd Density Physical Plane
Phi

Connecting the Dots

Let's look at Andrews recent contribution, connecting Phi with numerology, modulo 9 math function found in Rodin math.

We have Phi, 1, 1 / Phi, and 1 / Phi ^{2} as our current density model having 4 squares from the chart as one complete turn back on itself and inwards, one cycle.

The Cartesian formula for a circle is $x^2 + y^2 = a^2$. This is the pattern we see for the spherical Electron as well as the Atom setting in one density.

It represents the interaction of two of our three tempic vectors producing, Time, Electric, and Magnetic fields along the two tempic vectors as voltage and gravity which are the distance squared forces. This contains Pie or the Pi relationship and is the basic pattern of Electric and Gravitational force appearing in any one density.

If we now use Phi instead to relate a "cross density" field between two opposite sides of the spiral pattern:

As a round off, let Phi = 1.61803399
and let 1/Phi = .61803399

$x^2 + y^2 = a^2$
$Phi^2 + 1/Phi^2 = 3$
$(Phi^2 + 1/Phi^2)^2 = 9$
$2(Phi^2 + 1/Phi^2) = 6$

A beautiful mirror symmetrical power return to source as
$(Phi^{1/Phi}) * (1/Phi^{Phi}) = Phi$

The recognition of this number sequence 3 9 6 would seem almost too good to be true. This is the basis of the Rodin math principle, derived from the age old numerology system of modulo nine math. This is now related to Phi as well, or sums of Phi squared [Voltage and Gravity are the square of the tempic field in quadrature].

A Conceivable Application

From our models of tempic vectors forming Electric and Magnetic fields the following may become obvious.

3 represents the sum of two voltages in two densities from the lower density side.

9 represents the magnetic field resulting from both sharing one tempic vector down the spiral, yet creating a 4rth density awareness.

6 represents the two voltages interacting through the connection, or the doubling inherent in the path inwards on the higher density side.

9 becomes the connection where the energy crosses between 3 and 6 through a common magnetic field.
Since the spiral is progressive this same relationship should exist for all density thresholds.
Here in lies the 3 9 6 pattern from both ends of a magnetic field setting crossing between two densities.
The orientation of a magnetic field existing with one pole in each of two densities gives us a model for creating a flow of energy between densities.
This is the model presented with the Kosol Device [3SD], which has oppositely spinning spheres with magnets that cross allowing the magnetic field to form a pulse train where energy can move between alternate time frames. The very first 3SD had magnets in attraction, and this would create a small representation of what would happen. As the magnets cross, some form of energy would move through the connection related to two dimensions of the magnetic field in alternate time frames. This could be a voltage or it could manifest as a gravity field as both use only two tempic vectors in quadrature.
As we know, in a magnetic field that is in attraction, the lines of tempic spin fall into parallel travel, and in a Protonic magnetic field both orbital and particle tempic lines also attract. This creates 4 tempic lines all moving the same direction. If it is possible to place one pole of a Protonic magnetic field across a density threshold with all flux paths crossing the results may be a flow of energy outwards or inwards one direction across the threshold.

The model of the atom leads to the next logical step in realizing that the nucleus is exactly this. The Protons are already setting inside the strong force area where gravity is lower, their magnetic field reaches outwards to the Electrons magnetic field setting at a lower density. This model would seem to have application already inside the atomic structure as to how fields interact from inside to outside the strong force sphere.

In the case of a normal Electron generated magnetic field the flow would be bidirectional because of its dual flow structure.

It would appear that the Rodin math, based on numerology [3,9,6,9,3], may represent the functionality of energy crossing the density threshold through a single magnetic field that overlaps both densities. This gives us a mathematical model for energy moving between densities.

Don Mitchell's Dual Monopole Crystal Orb Pattern

Laying this model over the dual "like pole" crystal pattern suggests some interesting things.

We see two 3rd density magnetic fields crossing one another in a higher density at the equator where both share an opposite pole in another density.

His configuration is 3 at the poles, 9 as the next step inwards and 6 is the shared connection.

If this model is correct I would expect the field of 4rth density expansion to be strongest off the sides of the crystal Orb outwards from the 6 magnets that join both hemispheres, and inwards here as well.

Since Don is using Neo Magnets the device will flow energy both directions across the density threshold and a moving AC vibratory field would be able to cross density back and forth as an oscillation. However with the addition of the Copper wire or Aluminum, this may change as now an inductive metal is present. Because a Protonic magnet may move energy one direction depending on the field direction, this may have a new effect to channel actual power one direction. The power could be expected to manifest as either Voltage, Gravity effects, or both.

The Rain Maker Device

If the device becomes functional in two densities then we may be able to study the actual numbers. If the device crossed into 4rth density, then reversing the poles may possibly be used to bring it back, or it could be used to continue on, inwards to the next density without a reversal. This would probably require the use of the inductive metals.

The first notable sign of a successfully density shifting may be the center of the sphere, the equator, may begin to fade from view, as observed from outside the field.

This would also suggest that to simultaneously reach 5th density from here would require a larger form using one more set of transfer magnetic patterns. Another 3 9 6 9 3 stacked around it. This would suggest the second layer magnet structure setting around the first one and create two Density hops, however what pattern should be used may be up for some speculation at this point. If a single coherent magnetic field could reach through two densities simultaneously, then an AC signature could be transferred across both. Spheres laying closer to center would be expected to cross further as spheres laying outside would connect to this density.

This model also suggests that the proper places to tap a voltage from the system may lie between the equator and the ends of the sphere, using Bismuth, Aluminum, or Copper as active pickups. If these substances end up inside a canceling magnetic field they may begin to interact along the 90 degree direction and either become energized, or cold [cold electric current] along the equator of the sphere. The direction may be controlled by reversal of the coils polarity connections.

This also suggests a new model for tapping power of the alternate time flow rate from the 4rth density, using only one layer of magnetic structure, which seems to be a first.

This is of course one possible mathematical view of what may be happening.

On Density and Spiritual Planes

Also from the spiral diagram it is really unknown where exactly the next physical density lies. Whether it is half a spiral inwards or a complete turn is anyone's guess. In the case that it is actually a complete spiral cycle inwards, then it will take two layers just to reach the 4rth density. The spiral inwards is presently unexplored territory. The 3 9 6 model for X^2 force across density however now seems to lend a certain credibility to this model:

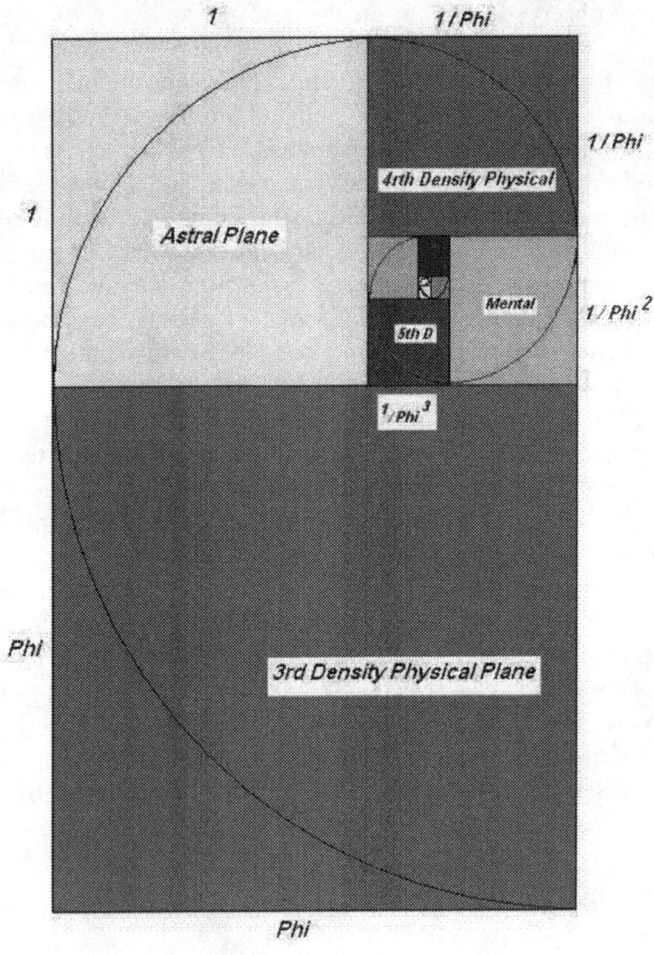

The Rain Maker Device

This chart shows the relationship of the Physical realms of Density as well as the spiritual realms setting in quadrature arrangement along the spiral path to Source.

We find the 90 degree magnetic field of the Smith coil causing human effects in both the Astral and Mental planes, which are setting 90 degrees to the 3rd and 4rth physical Densities.

This model can be used to relate the linear tempic differential, and the squared Voltage and Gravitational forces across densities as a Phi relationship.

A magnetic field with a single pole in one density and the opposite pole in the next density can be seen as a 3 9 6 relationship at the level of the area type forces.

Numerology uses this same 3 9 6 function probably relating to the Astral and Mental plane relationship.

As the distance from source decreases, time flow rate increases, gravity decreases, and space becomes more dense, or more energy in less physical space.
Matter vibrates at a higher rate, and to us in 3rd density 5th density looks like a world of powerful Celestial Light.

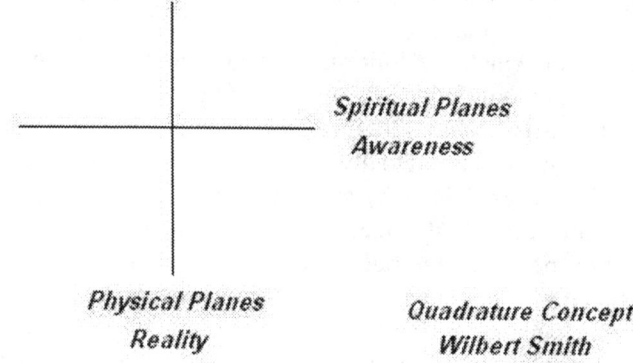

Spiritual Planes
Awareness

Physical Planes
Reality

Quadrature Concept
Wilbert Smith

It was Wilbert Smith who suggested that awareness or consciousness lies in quadrature to physical reality. Our experiments have led me to revisit this concept as both Astral and Mental phenomena have now been observed related to the magnetic fields produced in Wilbert's coil structures.

It is possible that all the planes in the Spherical universe model lie along the spiral in quadrature. That is the [Phi] square represents the 3rd density, the yellow box with edges of 1 is where we might expect to find the Astral or emotional plane. The [1/Phi] blue box becomes the 4rth Density and the [1/Phi^2] becomes the mental plane, [1/Phi^3] the 5th density and [1/Phi^4] the Causal plane etc. Physical planes located along one line and spiritual planes laying in quadrature to them and one layer higher in density.

This model pattern is suggested from direct experience in Astral travel where a new depth of dimension is realized on opening to each of the Spiritual planes of awareness. Interesting to note that possibly the same 3 9 6 relationship may hold true between the Astral and Mental planes and this could show us the substance of the numerology system for "life path" and the other intricate applications of this ancient system.

The model now suggests a device layout capable of tapping directly both the Astral and the Mental planes as well. It also brings a new meaning to the Rodin coil that might not be so evident. The nature of the quadrature magnetic field surrounding a Rodin coil may actually be placing fields into all four planes at once.

Two of which are Spiritual in nature and two of which are physical in nature. Since the Rodin donut pattern may be identical to the Electron orbital pattern then matter itself may be encoded with this same pattern actually spread over multi dimensions.

This also explains why so many who initially work with the scalar coils may first experience emotional effects when they encounter the 90 degree field and also very often may experience high speed mental comprehensions or downloads. It also confirms my own channeled or intuited information that Emotional healing or mastery of "mature emotions" is a necessary step to reaching a mental "comprehension." This represents the transition from the

The Rain Maker Device

Power center to the Heart center for the social consciousness of our planet earth as well as entering the upper chakra system represented in the upper tetrahedron of the MerKaBa system, the completion of the second half of the 3rd density, about to begin?

The spiral model of the Universe now becomes rather complete looking, including the physical worlds of manifestation as well as the spiritual worlds of intention and comprehension.

We have incorporated Wilbert Smiths Density model of spin in the physical realms, with the Phi spiral to Source [God Force] using a quadrature relationship. I have a feeling that Wilbert would have been rather impressed!

The Universe as Viewed from the Spherical Model

This model is one that was perceived during a meditation with a Crystal coil assembly and represents how the forces manifest in the spherical form to create the physical worlds or Densities. Although the word "dimensions" has been used to describe these separate worlds, they were previously discovered by Wilbert Smith during his research in the 50's.

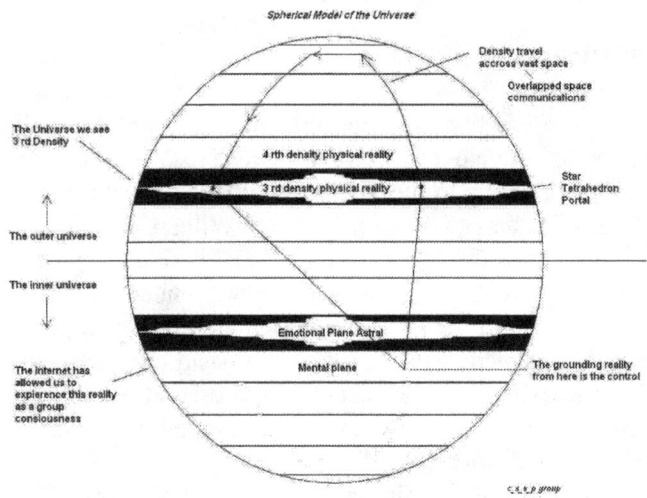

Notes

Phi = 1.61803399
1/Phi = .61803399
1/Phi^2 = .38196601
1/Phi^3 = .23606797

Phi^2 = 2.61803399
Phi^3 = 4.236067987
Phi^4 = 6.854101987

Round offs:

The spreadsheet figures are accurate to seven decimal places. We have loose ends after that (on the eight place, which I chose as the round off) because PHI is irrational.

Another 3:

Because of the unique inversion property, the sum of phi squares is the same as (PHI/phi)+(phi/PHI)=3

In Summary

The above is presented as a "possible model" to aid in the design parameters for device engineering of a "cross density" energy system. It is based upon the intuitions of Kosol Ouch on Phi and higher consciousness, the writings of Wilbert Smith for the tempic structure of the EM field and Quadrature concept, the Phi relationships presented by Andrew Bellon connecting Phi^2 with the 3 9 6 numerology system, the model presented by Don Mitchell of a magnetic Spherical crystal system crossing densities, and the applications, experiments, and observations by myself while working with the devices. The information on the Spiral and Phi is derived from a Wiki site on Phi.

A side effect of crossing a density threshold may also be a new awareness, or an expansion of conscious awareness of an existence much greater then our density. Once again we are seeing the integration of both the physical and spiritual realities combining to form a much more complete and all encompassing model of our universe. A bold endeavor started by Wilbert Smith 56 years ago, may he be remembered as one of the truly inspired seekers of truth.

Zero Point and the Five Sources of Magnetism

This paper was part of the development process which led to the final paper on the time vector model of magnetism. However due to some rather unique additional views on Zero point energy it is included here.

Electric Force

If we look at a wire with current flowing inside it one direction, we see a magnetic field coiling around it and expanding outwards many inches or even feet away. This magnetic field is much larger then the wire, or the current flow area. This tremendous expansion of the field may be from the two Electron magnetic forces actually pushing away from one another.

Looking closely at a flood of electrons traversing copper atoms all moving one direction we realize that as more cross the electron shells one direction we unbalance the magnetic field of the atoms. As this flow of electrons progresses from atom to atom, the actual electrons are moving around the atoms at near light speeds. Spinning the opposite direction that they are orbiting. The two magnetic fields produced are one inside the other, in opposition pushing time vectors towards one another. This creates repulsion between the flows and causes an expansion of the field.

The one field is from current flow through the orbital shell, and the other is a result of the atomic forces, and the natural electron motions as it orbits the atom with a reverse spin.

All force can be expressed as a time flow rate vector interaction. Time flow rate is far more powerful, and is present in all matter. Time flow rate is a function of "energy" moving in a circular motion at near light speeds. It is a linear force and the sum of all vectors reaching a single point in space. Time vectors originate within [matter] which is spinning energy.

The Rule of Time Vectors

Time flow rate vectors do not follow a summing in the traditional method of vector addition for solving for the resultant vectors. They follow different rules. Parallel vectors slow one another. Opposing vectors speed one another, and vectors moving away from one another do not affect one another. This rule is "intuitive" and found from observing magnetism at each level, the only method seen to make it possible for magnets to have both attraction and repulsion from the same pole.

This relationship is due to the nature of density. As time vectors collide density increases and time flow rate increases. This would seem to be the nature of hyper dimension. As time flow rate increases it hits a threshold where the pull reverses as it crosses the density threshold. The pull towards the "rest state" moves to a higher density, and begins to "accelerate" rather then to pull back towards zero point in this density.

Zero Point

The zero point of density 4 is a higher frequency and higher energy state then the zero point of density 3. Both are the rest states of one stability of matter, or energy in perpetual motion as atoms. We are all very familiar with the entropy of the 3rd density with respect to energy and matter. The zero point state of 3 density is not appealing, and we know intuitively that it is never an energy gain. Everything here slows down and stops. It winds down.

The zero point we need to approach to increase energy or to tap more energy is to reach out to the one that lies closer to center of spin. The next higher density. If an inwards link can be formed then the higher energy state can be used to flow into this one, and at the point of flow there will be an energy available that will far exceed anything we have touched before.

The Zero point is the point of stable atoms in a given density. They do not lack for power to exist. The atoms stability crosses the density threshold and moves probably all the way to [source]. Atomic particles are not always present, they appear only when they cross through our density to become pulses of force at specific times. Matter can be stable in any density. Density travel is a process of altering the matter and moving it to its next stable state.

A model for zero point energy must include the perspective of multiple zero points at the center of each physical density.

The Monopole

Proton magnetism is unique in that it results from both inside and outside movement the same direction. If it follows the above rules for time flow rate vectors then it may be possible to create a Protonic monopole magnetic field with its center flowing only outwards. Whatever state the nucleus of the atom holds would be coupled outwards with no resistance from an inwards flow. If we can extend the Protons magnetic field outside the atom keeping the opposite pole inside the atom pointing inwards to the Neutrons, we may discover the monopole in our density. This would manifest as time vectors moving away from one another from a center point, and having no countering force moving inwards to balance it. A magnetic explosion along one vector and yet keeping a compression along the other two.

This forms a picture of a strong force expansion, expanding as a torsion force where density is different inside and outside as within the atom, and as seen in UFO research.

The minimum configuration to accomplish a Proton monopole at one center point may be a platonic form. I would not be surprised to find this shape could be quite small, yet large enough to converge 20 or so Protonic magnetic fields. When the reverse poles flip inwards it may be enough to cross the barrier and tap a higher density with the other pole of the magnet, creating a vortex pulling one way outwards. Its North pole in this density and its South pole in the next higher one, and both flows coming outwards.

The Five

One of the messages I received early on in my outerworld communications was the importance of identifying "the five". It was just realized that my current model of magnetism is missing one important source often overlooked because of its thought to be weak magnetic properties. The Neutron. However the Neutrons magnetic field is present and almost as powerful as the Protons. It holds the form of Electrons field in that it spins opposite as Proton. Its magnetic precession frequency is high and may indicate very high time flow rates.

The five sources of magnetic fields are now identified.

Electron Orbital Spin
Electron Particle Spin
Proton Orbital Spin
Proton Particle Spin
Neutron Particle Spin

In the case of Electrons the two fields generated repel and balloon outwards. In the case of the Proton generated fields they are attractive and probably pull inwards. In the case of Neutron, this is the last remaining traces of Electron as it is almost completely neutralized by Proton in the merging. Its field remains Electron oriented and probably contains an equivalent magnitude or the Protons filed only in reverse, so it would offer the most inner repelling force at the atoms core.
    The Proton and Neutron lay within the strong force area in an altered density where gravity is slightly lower.

A Model of the Human Life Form

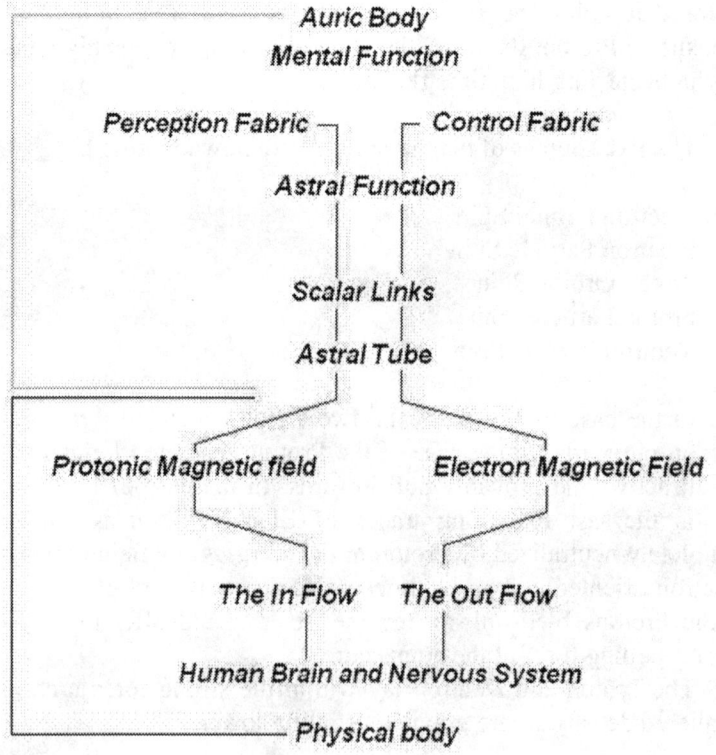

Wilbert Smith left a road map for those of us who have tried to follow and learn. The coil design he left us, and the quadrature concept, supposedly given to us by alien contact back in the 50's.

It makes sense that this is the device that would have been given first from an alien culture, as comprehending it's method of operation leads to an awakening process and a tapping of the inner perception fabric. This leads one to an expanded perception field and greater understanding of the human existence. When a culture demonstrates the ability to make the coil functional then it has shown its ability to turn inwards. Now ready for the transition to learning awareness and control of the emotional realm, rather then setting at its mercy in ignorance.

With our now present meditative abilities, available to any and all who are ready to look closely and begin experiment in this direction, we now have an added advantage. Today we have many good and solid inner awareness "mental models" to help comprehend and directly experience how we operate, so that soul can begin to identify itself and the relationship we have to both awareness and the physical life form we occupy.

Awareness Lies in Quadrature to Physical Reality

Although Wilbert's description of this is somewhat vague it is evident that "the boys upstairs" believe that this connection between conscious awareness and the physical reality is a quadrature arrangement. This means that as with all of the forces in quadrature they are actually "forces" that do in fact connect at 90 degrees to one another, are interlinked, and cannot exist independently. They are real, and we need to take notice of this new found reality to understand ourselves as well as how the physical universe works. This is the one thing that was impressed upon me during experiment 6 and 7. The reality that consciousness moves along the scalar wave form in and out of the fabric of awareness. Although the hard core scientist may still be inclined to state emphatically that awareness and consciousness are merely what we experience inside our brains, and that my experiences have been all inside my own head, I am convinced otherwise. I truly believe that my higher consciousness lies in a quadrature force field which has been called the astral plane and the mental plane by meditators, and I actually only peer outwards through the physical body. As the Spiritualists have stated, I am a Spirit Soul that has a body, not a body that has a Spirit.

Since this force field is real and it is shown to directly link through the scalar interactions of the magnetic field moving through both Electron and Proton vortexes, I believe any OU device should be able to bring some model of this into the system. This may be the piece we have been missing and why so few succeed at building a device that is functional.

The Perception Fabric

I have traced the perception fabric inwards, and indeed from experiment 1 on it has been present in all the projects. The conscious perception appears to move across the scalar forces present in the Nuclear vortexes of the atom. It uses the strong inflow energy of Source to transmit itself between the physical brain and nerves into the field of the perception fabric lying on the astral and mental planes of awareness.

The perception fabric seems to be "aware" of all the information, the forces, the flows, and the connections. It seems to know intuitively how they operate.

Following the nuclear magnetic materials inwards, [bismuth, aluminum, copper] by using them to generate a scalar or torsion field, allows us to establish a synthetic link. Our awareness may travel across them and experience the forces directly, because it is already adept at doing this the other direction outwards into our physical form.

This suggests that as we stand in the compressing field from the Cathedral Rock vortex in Sedona Arizona, we are in fact able to move our consciousness down and into the earth's levels of awareness. One of the other fields I accessed early on with the perception fabric was the Akashic record storage, the myriads of past events stored on the causal plane of awareness. I believe that it is all in there and may be accessed as the high speed download. Indeed as we progress along the spiral all the emotional right brain functions become finer and more pure. The Love approaches God Love and the higher vibrations, and the left brain mental focus becomes more pure as well. At each turn around the spiral we move closer to source, and each level must be also connected to the flows from source.

The Three Dimensional Phi Spiral

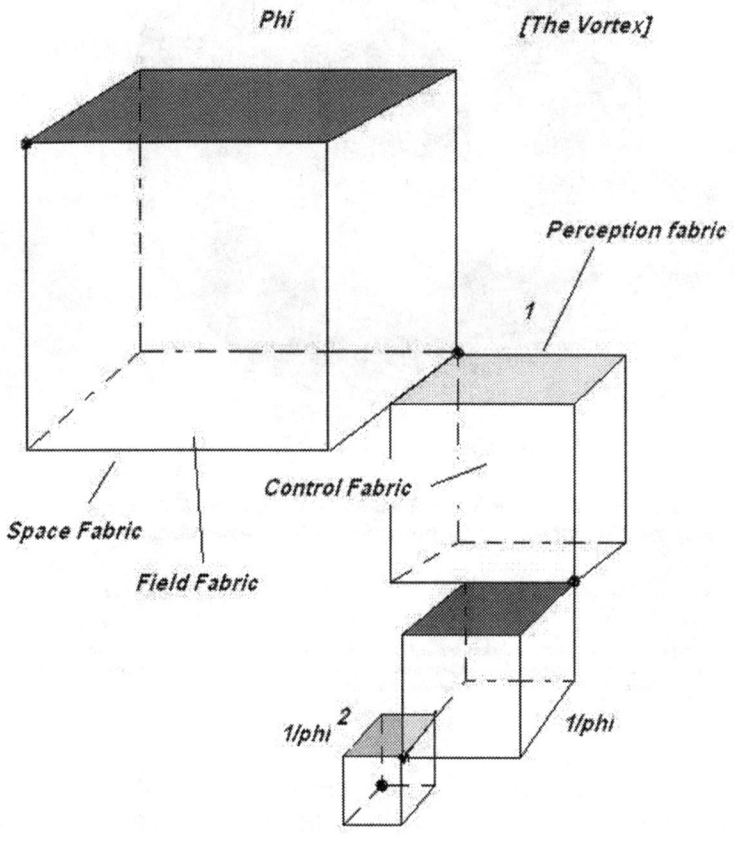

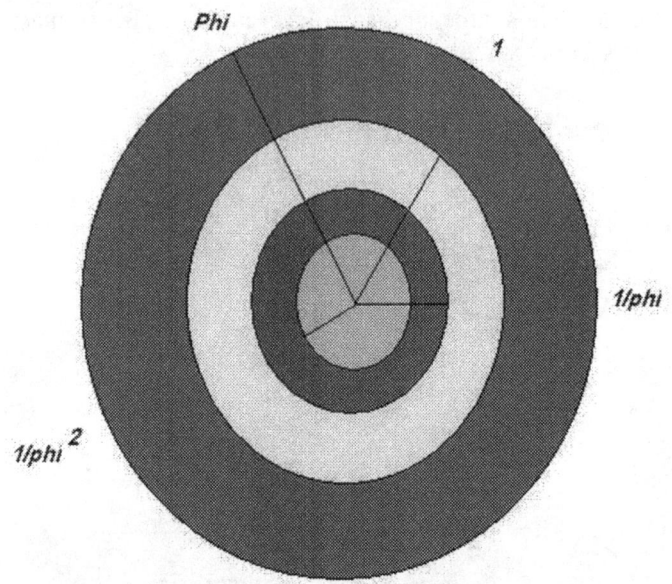

Density Model showing radius = Phi, 1, 1/phi, etc.......

The Area of each circle is made identical by altering the time intervals along the circumference for the spin model of each density

Wilbert Smith

Using Wilberts formula, we find the ratio of timeflow rate from one density to another based on the circumference being enlarged such that the area in each circle remains constant. In this fashion the time flow rate is found to increase as we move inwards towards a higher density.

From the chart we gain additional insight. As we move inwards to the higher planes such as the causal plane the frequency increases. A down flow from the mental plane [light blue] is faster then a download from the astral plane [yellow].

The Control Fabric

The control fabric is experienced directly through the Electron layer, the outflow, and is done by using the electron layer to generate a modulated magnetic field rather then the scalar Proton

The Rain Maker Device

layer. Magnetic materials like Iron that have little or no nuclear magnet involvement, and Silver wire to activate the pure outflow. As we wrap scalar coils on diamagnetic materials, conversely we modulate normal coils on magnetic materials to experience the outflow. Even a very small 2 volt modulation at 100 KHz wrapped over a strong Neo magnet allows one to feel the outflow presence.

Voluntary Verses Involuntary Control Functions

It should be obvious that some of the things we experience in life seem beyond our control. We are all here generating one reality, however few if any are aware that we are playing some part in the whole of the creation. My own personal model is a "cloud" of souls joined by astral cords in the body of mankind. We all can pull on the "cloud" a direction, we can all fight with it or go along with it, we have free choice. Few have actually "felt" it or learned to operate by using these connections from the inner awareness as I have discovered to be very real. Especially in the areas of emotion we are able to manifest relationships very quickly using these inner methods. It becomes obvious to any who go here that we are all in fact connected on this emotional layer and can affect one another tremendously at this level. The connections are real.

Some of the involuntary functions of the control fabric we really do not want altered, like DNA structure, the contents of a grain of corn, things we take for granted but really do not want to have to keep thinking about.

So the control fabric at our level of "operating" is limited to the planes of awareness we are aware of and have learned to use effectively.

I would assume that on some deep level all mater is being operated in this fashion, however few if any may ever become aware enough to actually begin effecting matter and altering its form at will. The intuitive skills necessary would be very advanced indeed and lie much deeper then the astral plane of awareness.

Realization

When I had the device of experiment 7 set up and had begun working with trying to develop a loop function of outflow to inflow, came the realization that I already exist in the inner planes of awareness and the "inflow device" should not be necessary for me to manifest an interaction into the outflow device.

I noticed an intimate coupling to the device I had constructed, and to all the basic principles it uses to function. I had sent vibrations through this loop and felt them become amplified. This next step was a leap indeed, like waking up from a long dream. I started to "feel" the device from the inside outwards and began to manifest the emotional energies directly from the inner planes of communication. I then took my hand away and started using the device as a psi activated device from several feet away. More tests must be done along these lines but I believe that the device can be operated from the control fabric directly and no physical contact is necessary to transmit "emotions" through it to others. This includes the healing energies and all the vibrations of "perception" including telepathy with pictures colors and words. All information should be able to pass through the control layer outwards to a person receiving at the device. In this way we can determine the distances and speeds of the control fabric, but more importantly we can learn to "feel", use, and experience it first hand.

The Rain Maker Device

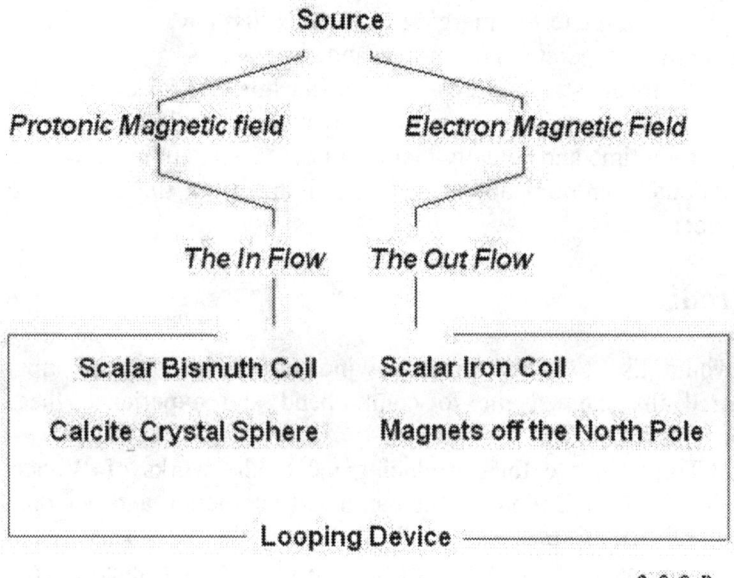

In Conclusion

It was discussed at c_s_s_p that sooner or later we will probably discover that our human bodies are in fact the ultimate tools for interacting with these "fabrics" of reality, however these experiments could begin to lay a basis for mainstream science having to finally embrace the idea that consciousness may in fact be located outside the humans physical form, linked to it by the forces of nature that until now have been hidden in their methods of interaction and reserved for only the experienced meditator of our present day "cloud". Experienced as the auric layers that sensitives state do exist and can be sensed, now referred to as the fabrics of perception and control, the "aura", actual fields of force connected to all physical matter.

    Also it was noted that physical death would lead to the withdrawal of the sensory and control fabric link with the physical body, but very probably not the end of the personal soul's existence. This puts a new pressure on the religious community as well as the scientific to continue to search out and begin to complete the merging of models that have up until now been totally separate in our minds and experiences.

As to ghosts, it is possible that an inner astral or departed soul could in fact share a part of anyone's control or perception flow for a time and actually manifest both visual, audio, as well as emotional communications with the living using these fields to interact.

Credits

I would like to thank the following participants for the input shared that helped me to comprehend and experience these concepts first hand.

Trent George for introducing me to the works of Wilbert Smith. Andrew Bellon for his continued interaction and personal comprehension process, Dell Coleman for his very keen observations as to the nature of the conscious energy encountered going inwards, as well as all the participants of the group at c_s_s_p and the group at Kosol_Core_Tech for their presence. Also Don Mitchell personally for the tremendous support offered in this inwards searching process, his tremendous wealth of knowledge and effort that has been offered freely and openly, shared in a spirit of common good for mankind. And Kosol Ouch for his ever present and direct lessons on the inner planes of awareness.

I believe that although sometime individuals may awaken in an instant, the collective of mankind moves in tiny baby steps, very slowly, but never turns back once truth is realized by all.

Time Vector Model of Magnetism and the Aether

All the forces can be represented as linear [inverse distance] Tempic field interactions crossing one another in quadrature.

A "tempic field" or time vector is perceived as the basic structure of the universe and is always found moving in a circle, which is the only structure that all awareness can agree upon. [Wilbert Smith]

The picture of a magnet shows flux leaving the North pole of a magnet expanding out and up then taking a slow curved path all the way around to the bottom of the magnet. The ends of the magnet contain the highest intensity of flux per square inch, and as we move along the flux lines they balloon outwards as they curve back on the magnet, then re compressing into the other end.

If we picture each flux line as a coil containing one "spin" vector that moves forwards while it circles, this forms a mental picture of what a "spin" vector would look like which originates within a circular atomic motion. A long coil or spring shape running North to South with an arrow in the Z axis to show the direction of advancement along the coil while the spin vector winds it's way along the other two directions of motion X & Y. These X, Y, and Z vectors are the time vector component physical equivalent vectors in 3 space for quadrature interactions.

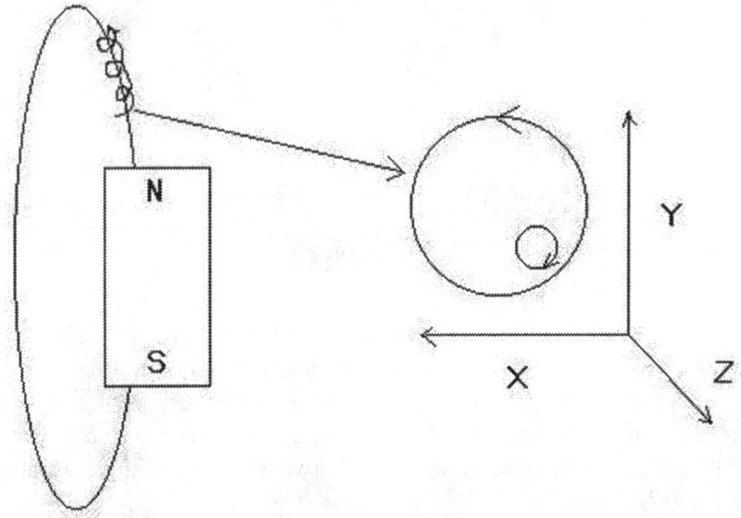

Now picture a close up of this "spin" vector or spiral, sliced and brought outwards to inspect, as in the diagram above. As the spiral circles in physical space it moves through two dimensions, X and Y. As it advances it moves through another the Z axis which curves along the flux path. These are the "time" vectors of the magnetic field or the "tempic" vectors components of motion X Y and Z. All three must be considered when determining the interaction of mixing the 5 fields of magnetism.

The Electric and Magnetic Fields

The circle $[x^2 + y^2 = r^2]$ built around two X and Y spatial dimensions, and represents a voltage or static charge. Its force falls off as a function of inverse distance squared. Although it may be true that as any two tempic vectors cross at 90 degrees there is a voltage component present. Voltage is the first x^2 force to be recognized as moving outwards from creation by Wilbert Smiths model. It always seems to appear in the form of a circle or the surface of a sphere.

Adding a third tempic vector of motion Z creates the B field or the magnetic field and the equation now becomes a third degree equation representing a volume the shape of a donut. The

The Rain Maker Device

three component vector tempic fields are now setting in "quadrature" or all have 90 degree relationships between them at any one point. Adding the third tempic vector of motion creates a field that curves back on itself. This would seem to be the limits of rectilinear space for a 3 dimension system.

The Tempic field drops off as a function of inverse distance and in this sense is linear.

The Electric field drops off as a function of inverse distance squared because two tempic vectors are required to construct it.

The Magnetic field drops off as a function of inverse distance cubed because it is a volume or three tempic vector components.

These are the appearance of the first three forces: Time, Electric, and Magnetic fields. The nature of Tempic vectors interacting in quadrature.

The Five Fields of Magnetism

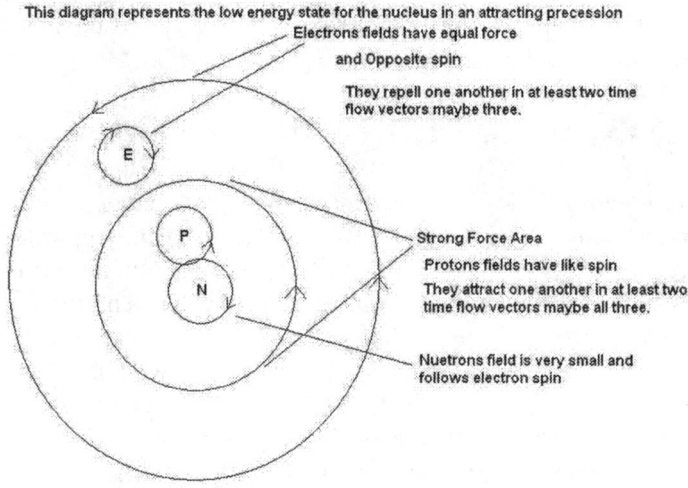

This diagram represents the low energy state for the nucleus in an attracting precession
Electrons fields have equal force and Opposite spin
They repell one another in at least two time flow vectors maybe three.

Strong Force Area
Protons fields have like spin
They attract one another in at least two time flow vectors maybe all three.

Nuetrons field is very small and follows electron spin

The Protons magnetic fields radiate from a different density, one of lowered gravity.
These are the five possible components in a complex magnetic field.
Magnetic fields balloon up far in excess of the electron physical flow circle. Proton magnetic fields may contract inwards instead.

A list of the five component field sources for tempic spin:

1 - Electron Particle Spin
2 - Electron Orbital Spin
3 - Proton Particle Spin
4 - Proton Orbital Spin
5 - Neutron Particle Spin

Magnetic Fields Interacting

A magnetic field can be compressed or expanded along any one of the "time" vectors depending on the presence of another flux spin vector passing by it or through it inside or outside. The spin vector rotates along all the time vectors of 3 space. The "spin" vector represents the motion contained in the particle or orbital motion responsible for the magnetic field. As any one tempic vector is altered, the magnetic field will compensate to stay in this density or local time flow rate, by expanding or contracting it's other two tempic vectors in rectilinear space. This is the observed nature of magnetic fields.

The Expanding Electron Magnetic Field

The Electron field is composed of two spin vectors, or coils of tempic field. The Electron particle is spinning the opposite coil direction as its orbital spin. The Electrons magnetic field is donut shaped and expands outward many times larger then the magnets physical size. This expanding property is one that has been a mystery. I theorize that the two spin vectors or tempic fields, also advance the opposite direction through one another and this explains the Electron magnetic fields expansion along two of its time vectors setting in motional opposition. Even though the two tempic fields originate within the motion of one particle, the Electron, as they leave the top of the magnet they separate due to the opposing forces of the component tempic vectors of motion. This creates the familiar large ballooning donut shaped field.

The Contracting Proton Magnetic Field

The Protons magnetic field is composed of two spin vectors, or coils as well. They spin the same direction. Both spin vectors travel almost parallel, only the inner one is spiraling as it advances inside the outer one. I theorize that the two spin vectors advance the same direction through one another and this should result in a contracting in between the two coils due to tempic or time interactions along all three of the time vectors X, Y, Z components. The model that produces this concept is based on how tempic vectors interacting affect one another to form the forces of attraction and repulsion.

Deciphering the Interaction of the Time Vectors

There is only one obvious model I can conceive that determines how time vectors interact that can explain a magnet. There may be others that I have not seen.

A magnet repels like poles. A magnet attracts opposite poles. This can not be explained with only one spin vector. A spin vector only has one advancing face so could only either push or pull from one end but not both. Observing the expansion of the Magnets flux path and its attraction at both poles here is the model I have formulated. There are two directions of advancement in a magnet for tempic fields lying along the flux path. They move opposite directions one inside the other. Because they advance opposite directions they also repel one another creating an expansion of the orbital flux field, and a compression of the particle flux field deep inside it.

Because they spin or circle opposite directions, one can be interacting to attract and one neutral at the same time. In order to explain a magnet we must decipher the time vector interactions. This is the key to making the tempic vector model work.

The Time Vectors in a Magnetic Field

1 - Time vectors moving towards one another [add] to speed the time flow rate, this creates repulsion, as matter always moves towards a slower time frame. This is the model of gravity as well for time frames. [Einstein ... time flows slower near a strong gravity field]

2 - Time vectors moving in parallel slow one another, and slow the time flow rate, this creates attraction, as matter always moves towards a slower time frame. This is seen from observing the resulting parallel vectors found in magnets attracting along the attracting faces.

3 - Time vectors moving away from one another do nothing and are ignored generating no force. This is the only perceived way that poles can interact as they do and yet little supporting evidence exists to prove this at present.

Now if we apply this model to the Electron model above we come up with the magnetic field that we observe in normal magnets.

Two North or South Poles Facing Repel:

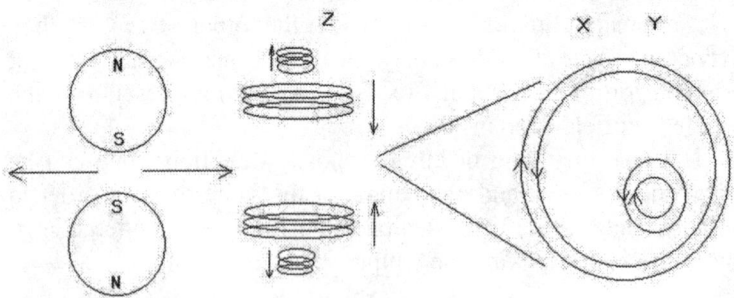

The advancing time vectors of one motion [orbital?] move straight at one another creating a faster time flow rate resulting in a repulsing force. This causes an expansion of one time vector

The Rain Maker Device

and bends the opposing fields out a great distance as they compress along the meeting place between the magnets.

The withdrawing time vectors of the other motion [Particle?} move straight away and are ignored however they are isolated now somewhat and do not fully complete their circulation.

Now viewing the circular or remaining two time vectors, they are moving in opposition along both directions of the circle. This causes the two fields to create between them a faster time flow rate resulting in repulsion along these two time vectors of 3 space.

The particle magnetic field is compressed inwards and the orbital field is expanded outwards causing the field to expand along two dimensions as it leaves the magnet. However since one of these flows is weakened at the repelling end, its return flow is caught in the moving away while being stuck inside the other field we may expect the balloon effect to be lessened along the distant pole and the field would extend much further backwards. This is in fact what happens in a tripole arrangement of magnets. The field balloons out in the opposition gap and it extends straight out the other end with less widening.

A North Facing a South Pole Attracts:

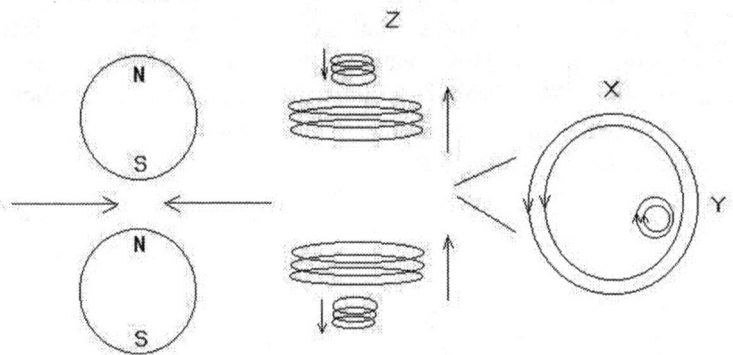

All advancing time vectors meet and travel through in a parallel motion of advancement. One spin vector moving the opposite direction through the other.

Parallel motions create a slower time flow rate and result in attraction on both flows running through the poles.

Because both flows still repel along two time vectors, the two tempic fields still balloon outwards around the magnets only now forming one large field moving through both magnets creating the familiar donut pattern.

This model offers an explanation for the entire magnetic field we are familiar with as originating from the Electron Shell motions.

Proton Fields

The Proton magnetic fields are present in the materials Bismuth, Aluminum, and Copper. Although they do not usually extend outwards of the atom it is theorized they may be extending outwards or at least present in the 90 degree field of the Smith coil, as Electrons fields are contracted inwards or canceled. The 90 degree offset field that appears may contain a component of the Protonic magnetic field. The 90 degree field has also been identified with an Astral or Mental plane connection through direct experiments.

With Proton magnetic fields it is unknown if the chart pictured above is accurate or if the Z vectors should actually be reversed. However it is known that the X and Y vectors do in fact run in parallel paths. This creates an attraction between the two fields resulting from Orbital and Particle motions and as the field projects outwards from its source point it would not be expected to swell or balloon as the Electron generated fields do.

The Rain Maker Device

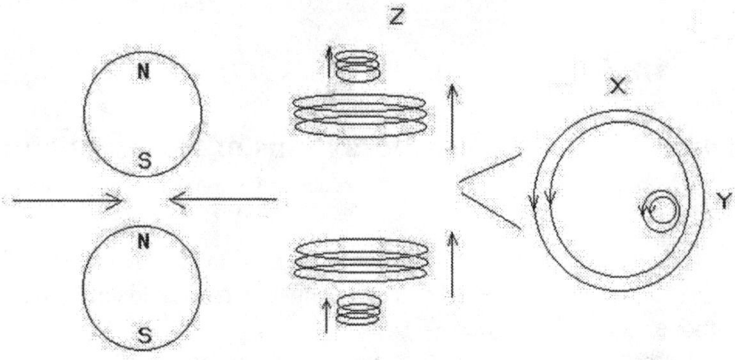

If the above model is accurate then the possibility of monopole magnetic fields should become obvious. Or at the lease if we orientate two Proton magnetic fields correctly there may be no interaction between them as all tempic vectors are moving away from one another. Reversing both poles we would then see repulsion in only one configuration.

Indeed a "repulsion" seems to be the only interaction we get when bringing magnets near spinning copper cylinders or in bringing strong magnets near bismuth.

The Protonic field seems to offer a one directional tempic vector field where all tempic lines are moving in parallel.

This creates an interesting situation since we find the Proton setting within the strong force area of the atom in the lower gravity area.

Since its field connects with the Electrons field, it becomes a magnetic field moving between two densities.

This sets the stage for a magnetic field that can flow energy one direction across a Density threshold and is modeled in the section on Phi and density.

Two or more Protonic magnetic fields setting in this configuration could form a monopole in this density.

Section 2

Proton and Electron Interactions of the Magnetic Field and the Aether

The Rain Maker 1 device supplied information on the nature of the Proton magnetic field by separating the two fields into two components of one device.

An outer ferrite iron ring and a central Bismuth core. An external magnetic field around the iron cylinder is set up using powerful Neo magnets with North in or South in and the interaction is observed to create either an outflow or an inflow of what has been called chi energy, a force operating in the Aether, or as Wilbert Smith calls them the control and perception fabrics of consciousness.

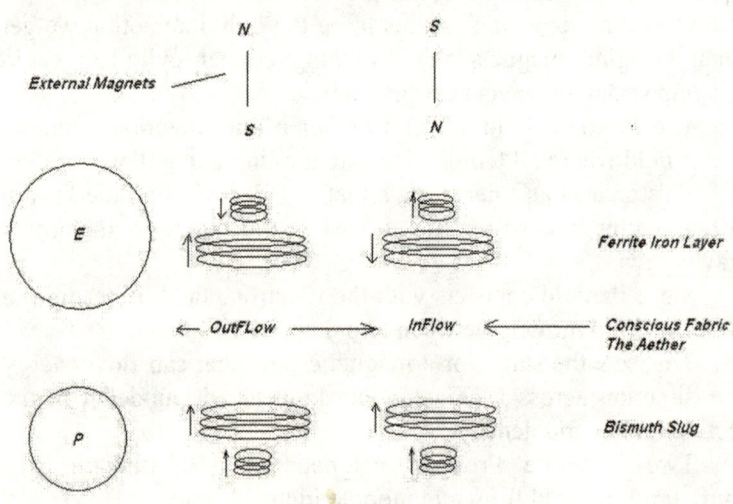

The Protons field always turns towards the Electron magnetic field when present and creates overall magnetic repulsion from either polarity.

However the Conscious fabric at 90 degrees to the B field receives a signature expressed as outflow or inflow and is a scalar

The Rain Maker Device

force that can be sensed with the hands as either a hot outflow, or a sucking cooling inflow.

North pole of the outer external magnets are pointing outwards on the ring. The Tempic Vector model of magnetism shows particle spins repel, and orbital spins attract. We now have a compressing magnetic orbital field setting around an expanding magnetic field from the particle level. This is the formula for outflow energy at 90 degrees to the B field.

South pole of the magnets are now pointing outwards on the ring. The Tempic Vector model shows particle fields are contracting in and the orbital field is expanding outwards. This is the formula for an inflow energy at 90 degrees to the B field.

Since both orbital and particle magnetic flows are always interacting between Electron and Proton, we can expect the interaction to be stronger then with normal magnets having only Electron Electron field interactions where the ones traveling directly apart do nothing.

From the above we may see that the perception and control conscious fabrics lay at 90 degrees to the magnetic vectors and operate from the scalar compression and expansion of the two magnetic fields between Electron and Proton.

When the two flows compress together we get outflow, when the two flows pull apart we get inflow.

This is all a result of the 3 tempic vectors acting on one another in the magnetic field between Electron and Proton.

The magnetic field between Electron and Proton act directly on the conscious fabric in this way. They actually set up compression and expansion differentials in the Aether fabric which if set off balance and then looped remain fixed there.

This explains why the energy feels radiant through the entire field and not alternating along the field from inflow to outflow. The entire magnetic field of Rain Maker 1 is either outflow or inflow but not both. This field radiates off the magnetic field at 90 degrees to it in a spherical fashion and the corona gets very hot to the hands.

The Aether and the Magnetic Field

"Tubes of force" [Tesla]
"Aether" = "The Medium" [Tesla]

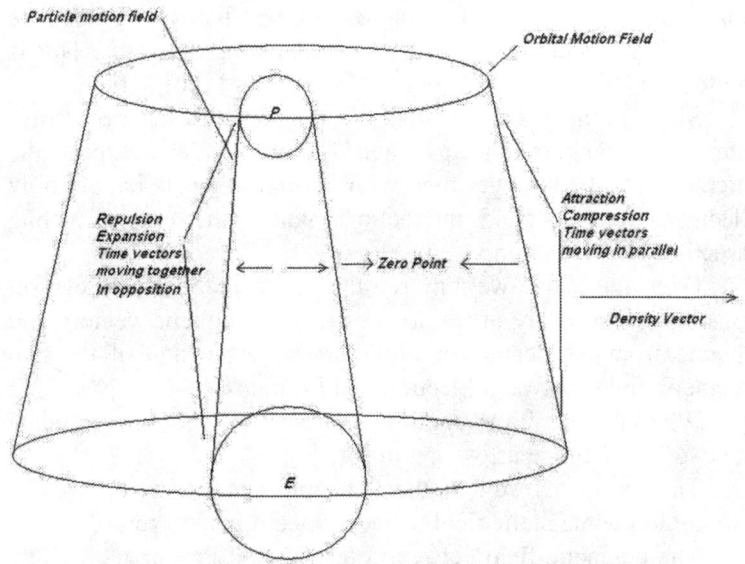

My observations with the Rain Maker 1 experiment led me to the unavoidable realization that we need some form of Aether model for engineering devices.

I observed a field that exists independent of the device once removed from the place it sat for a week creating the field in the background Zero Point.

The Rain Maker Device

The change in the background Zero Point level was observed by Tesla as well and he referred to them a "tubes of force" in the "Medium". His model of Gravity must have been from observing the same basic realities I have now seen operating between Proton and Electron magnetism.

Once the background "medium" has been altered it seems to store the [charge]. According to my theories this accompanies an altered Time Flow Rate in the background Aether as well as an altered Gravity in the area affected. This has yet to be tested using an accurate spring type scale rather then a balance scale.

Experiment 7 was not a welcomed event in my life, but the result is unavoidable at this point and leads to a new model of density including the Aether.

Proton and Electron correspond to the yin and yang forces in this density, or 3rd density. They are the opposite energy sources, however higher densities lie not inside one or the other, but in quadrature along the Plane of Aether modulating in the magnetic field between them. The tubes of force are acting on the Aether directly, and thus consciousness. Rain Maker 1 sets the two fields into an interaction and pulls not one but both together towards a different density, depending on the magnetic field we place between them. North pole pointing away from Proton pulls both into a higher Time Flow Rate. Together they move along the quadrature scalar force of the Density Vector.

The effects of Gravity and Time Flow Rate, as with Voltage, always appear in a spherical field. This would seem to indicate that as we set up a Gravity or time effect it will always manifest a spherical effect. This also was observed in Rain Maker 1.

Tesla and Using High Voltages

If voltages are to affect Gravity they will have to appear in perfect balance, that is + 300 v with - 300 v to maintain the Aether's perfect scalar balance. Using the magnetic fields of Proton and Electron materials avoids this problem by invoking a natural interaction between both whereby the entire scalar field can "slide" along the density vector setting at 90 degrees to the B field.

Tesla clocked the transfer "medium" speed for cold electricity at 471,240 Km /sec and the speed for the Electron transfer medium speed at 300,000 Km / sec.

This also clarifies the longitudinal voltages to modulate the Aether directly in Tesla's work. The Proton energies travel through a different medium then the Electron energies. One that lies in a different density.

Using high voltages to alter the Aether is not necessary as we have shown, devices like the Searl disc and the Hamel cones also show this. A more natural method to get matter to slide along the "density vector" is with the balanced magnetic field or B field set up between Proton and Electron so that they themselves can be moved together in balance along with the tubes of force connecting them.

An observation also on the work of John Hutchisen using Tesla coils and RF to send objects through density in a rather violent method that can even fuse two pieces of matter together or break the crystal structures of metals down so that they fall apart. This may be the results we would expect to get from manipulating only the Electric fields and not counterbalancing the Protonic forces along with them through the magnetic field. John's generators all affect only the Electron side of the balance, the yang force.

The Slow Building Effect of Aether Pumping

We have observed the "delayed" effect of altering the Aether, or the background pressure in the "medium".

It would appear that if we are to maintain the coherence of matter as we push it towards higher density, it may be necessary to work with the "delayed" effect to discover how to increase its speed, how to stop it, and how to reverse it. The delayed resulting Zero Point offset is altering matter at the base level between all particles and forces connecting them, and each must come up to speed together. It is also altering the Aether fabric as well and this must be studied further as we have no desire to alter the entire earth's space by the presence of many such devices producing power and transport. If the "tubes of force" do in fact appear in the background Aether even when matter is not present, then we may wish to become aware of them and even devise ways to measure them or even draw power directly from them.

The Magnetic Field and the Scalar Tube Forces

In my "vortex analysis" I theorized the following relationships are suggested.

Zero Point - Gravity = 1 / Time Flow Rate^2
ZP - G = 1 / TFR^2

This model allows for a negative or positive gravity.

I jokingly refer to this as the Lowrance equation however others have come up with similar descriptions more adept at math and physics. However even Wilbert Smith did not indicate a positive and negative gravity was possible because his formula was based directly on a squared function not allowing for this.

Along the magnetic spin tubes connecting Proton and Electron we find that gravity is altered as we slide down them towards Proton. Proton sits in the strong force area with a slightly lower gravity then Electron. This is the only formal appearance of an altered gravity field in modern physics although it is often overlooked because it is presented as "mass" rather then "mass attraction" or "gravity" which is not understood at present, so to be avoided at all costs.

It is imperative to realize the nature of the Aether background as a Scalar canceling force in 3rd density. It operates like charging a capacitor, only instead of voltage we are charging the Aether with [plus or minus] Gravity. On inspecting the graphic above, the Aether is altered in the interaction between the opposite magnetic forces affected by the time flow rates in two dimensions of the fields. Gravity is a two dimensional manifestation [distance squared force]. It is the two dimensions lying perpendicular to the B field being compressed and expanded against one another. While the standard EM fields we are familiar with set in quadrature are merely Electron magnetic fields interacting and give us the Electronics theories found today, they have missed the deeper interaction of the inductive metals and not made the connection to Gravity as a force setting in the same plane of motion but inside the Aether rather then in normal space or the field fabric and operating between Proton and Electron. Gravity is setting in the field fabric of the next reality along the spiral. Voltage is setting in this one. Thoughts of "Aether pumping" now are explained as we see the charging of the Gravity scalar is a function of two opposing Voltage scalars in canceling configuration.

Since the voltage and gravity scalars are both tempic squared forces laying in the same plane of alignment it is easy to see the confusion, and some have stated that gravity is voltage. They have missed the fact that gravity is linked to "time flow rate" and voltage is not. The compression between Proton and Electron magnetism is the key to pumping the Aether, and altering the gravity scalar which is not fixed, and its value being squared is altered with it's time vector.

To alter the time vector of an electron voltage the spherical model was offered above, however this method will probably not produce a density sliding effect and this is the difference between Tesla's' longitudinal generator and the Searl disc. To mimic nature we must simulate the Electron in motion, not stationary as a field force without its Proton counterpart. Until which time we have mastered "cold electric current" the longitudinal spherical generator is not feasible to me as a true density device although it may very well alter gravity, the results would not be balanced unless a cold electric field could be set up on the inner sphere rather then merely a positive voltage.

The Rain Maker Device

Working With Conscious Energy

In the event that some are new to working with conscious energy, or have never encountered the healing arts or therapy, there is a standard of moral suggestions that may help to ease the transition into becoming attuned to the flows. It is my suggestion that everyone at some point pursue their own emotional healing process.

In the mean time there are suggestions that may help avoid some of the pitfalls encountered when opening the connection to the layers of the perception and control fabric that lie outside the physical body and inside the aura. The following was offered by volunteers who have been testing the energy of the Rain Maker and Elf devices configured as an outflow device, as well as my own comments on finding a higher moral perspective and using the energy.

Introduction to the Process of Auric Awareness

One of the first comments I get when I throw the on switch to the scalar coils is, "I feel slightly distant from my body, as though I were outside looking in." I explain this is the first sensation in sliding. One instantly becomes aware that they exist somewhere else other then only inside their physical body. It is essential to realize this is merely the focused awareness becoming attuned to the existence of auric reality. If given this explanation people seem to do well with the experience.

The next thing is to realize that stored in the auric layers are often many trauma events, and if they surface it is necessary to learn to always turn directly towards the fears and heal them with Love and rational mental reasonings. However, this choice must be free to the individual and never forced, only offered. Many times a person is not ready to open an old emotional wound to reveal its true contents and will insist on leaving it shut at the cost of memory loss during the time frame of the trauma event. This individual choice must always be honored.

The next very major issue is that of privacy. With a device that opens two people to both auras contents, there is no way to lie or hide from what is present and available. We must all learn to never reveal what we have seen in someone else's aura. Ignoring this warning brings direct karmic involvement of you into their karma. You may be able to lift their burden only at a great cost to yourself. Never try to fix someone or force their "process", allow them to become aware enough to heal them self.

Let go of all judgments, biases, and prejudices. What ever is in someone else's aura is theirs personally. It has no reflection on you, your self worth, or your relationship with them. If you do not follow this advice you may become locked into an astral battle with them if you are perceived as the enemy. The nature of the bismuth coil operating with a sine wave should act to bring the mental plane awareness in as well and this situation may not be likely, but it is always good advice if doing any healing work with someone else where the higher perception layer is involved.

Initial Contact

The first interaction with a conscious energy device will probably be somewhat amazing, although it may take a little time to become aware enough to fully sense and use the energy. Allow ample relaxed time and conversation to slowly become aware. If attention is turned away for a time the connection drops away.

Pointing a palm at the device tends to raise the energy level, and thinking about drawing energy quickly produces a strong flow.

The meditative state is much faster because the awareness is already shifted into the auric layers and there is little clearing and focusing time necessary.

As long as the device offers a strong outflow there is really no danger of being sucked into the Vortex. Entering the Vortex would be possible if the magnets were all reversed to North inwards direction, the bismuth coil was turned to max output, and the ferrite coil was shut down. A strong inflow like this can be used to access the Vortex, but it is recommended strongly that both astral and mental connections be firmly established to prevent the discomfort and disorientation of the "slide" inwards, as well as a warning for everyone to take their raincoats if they go outside. A sleep state may also result. I have done this at least three times and noticed a loss of physical energy on every trip inwards. Keeping the vortex generator in the outflow configuration always seems to provide the expanding field and raises the person's physical energy state.

The Nature of Conscious Energy

The energy of outflow and inflow both seem to respond well to "intention." The flow can be accessed from the device drawn into the body and consciously projected to any part of the body or out the hands to others. It has been shown many times in our tests to follow the direct intention of anyone who has yet experimented with it. The more one works with it the faster and stronger it seems to flow.

If directing the focus towards the other planes of awareness such as the causal plane we have found access to the past life storage at high speed as a download.

This was only tried twice, both times with a minimal success because we are not yet adapted to receiving information at these high speeds.

The flashes from the causal plane seem to be very fast. This tends to indicate we are not limited to the astral and mental planes of awareness as originally assumed.

When locating acupressure points very little searching is necessary. The points pop out and can be felt inches away, easily zeroed in on and then seem to open quickly.

Future

I look forwards to following this line of experiment as time permits, however the original intention of my magnetism pursuit has been that of unlocking the key to opening the Zero Point Energy flow and achieving over unity power in a device. The healing aspect of these devices was unexpected yet recognized as very much worthy of documenting and developing to some extent. It is hoped that others will embrace this endeavor and see where it can lead.

Disclaimer

The energy devices presented in this book are for experimental purposes only and no claims of healing, either emotional or physical are made.

This energy is new and little is known at present, thus the need to experiment and keep the format in this mode.

Scalar waves at high levels could possibly be damaging [Philadelphia experiment], and it is not known what effects may result from exposure such as involuntary sliding or loss of reference and disorientation with the physical reality.

The bismuth coils alone have been known to cause temporary numbness in the nerves at high levels, but using them inside the ferrite shells has not produced any such effects we are aware of. The numbness has always faded in a matter of an hour or so, but one should be aware. The inflow energy can stay in the body for a time after the exposure, especially if the focus is put on it. If this occurs holding the North pole of strong neo magnets as well as breathing the energy out can help to release it, or even setting in the outflow field of an active device.

The Rain Maker Device

Vortex Analysis

"The Universe contains no anomalies, and the appearance of an anomaly is warning that our understanding is inadequate." Wilbert Smith.

Space is linear no matter where we investigate it, and this relationship must be accepted as fundamental.

When a meditator extends their hands over a crystal and feels a sensation, or seems to be communicating with a quartz crystal, we now have a good clue as to what it is they are feeling. It is a scalar energy that the crystals seem to receive and then amplify back to us. It cannot be measured with a compass or a meter in the conventional sense. Experiment 7 has taken this energy one step further and created a 3 foot field approximately 100 times stronger then any crystal I have ever felt, with sensations so hot they approached burning sensations and caused nerve tingles for a couple hours afterwards.

What the vortex has shown us is that we can set up an alternate time flow rate in an area of space that will flow along with the earth yet remain separated from it by a tempic corona sheer layer. By linking the Bismuth Proton time flow rate and the iron Electron time flow rate with a magnetic field and a scalar wave we have altered the time flow rate in an area of space to reflect something other then the ordinary time flow rate across this region of space. Along the corona of the sphere we created, was a hot sensation indicating a North out field was used to create it. With a South out field the time flow rate can be reversed creating the inflow.

By connecting the flows of Proton spin to Electron spin in two materials we have joined the energy across 3rd and 4th densities.

The time sheer corona was forming at approximately 3 feet across, not big enough to sit inside.

As the vortex resembles the form known as a "Consciousness Unit" or a CU, I may refer to it as a CU until this is proved one way or the other.

The fact that the sphere remained perfectly motionless leads to the following postulates.

Rules for Interacting With a CU

The following are intuitive observations and need to be verified by experiment:

Postulate 1

The CU will maintain whatever motion it had when it was formed.

There are three possible forms.

1 - stationary
2 - spinning
3 - accelerating

If true here in lies the key to power generation and stopping runaway devices.

Evidence of this lies in the fact that my first partial CU remained motionless.

Considering the earth is moving through space with spin, then it is very likely that the unit merely maintained the spin present when it was formed.

If we create one on a spinning device with a constant rate of motion the CU should be very stable in that motion and neither accelerate or slow down.

Postulate 2

When the CU is fully formed it will disconnect from the local fabric and become mobile, through a mental link?

Postulate 3

A spinning CU will remain linked to the machine that created it and continue to spin the machine along with it through the same fields that created it. Evidence is the Searl disk.

If it crosses the density threshold it will appear to accelerate away, but will actually be moving only through density as the space of the universe is smaller in a higher density it will move towards the center.

Uniting the Magnetic Field and the Conscious Fabric

Nuclear Versus Electron Magnetic Fields

Bismuth:

Coupled to both perception and control fabric and allows two way sliding if in the scalar configuration.

Magnetic field is active at the nucleus.

In a scalar coil the Proton field places an imbalance in the Electrons canceled magnetic field and causes one side to become dominant. So here we see that the inflow or outflow affects a normal magnetic field to make it stronger or weaker. The compass held around a scalar coil wound on an inductor shows this effect. Flowing DC through the coil produces a 90 degree field that reverses if the DC polarity is reversed.

It may be assumed from this that the Protons magnetic field then contains the outflow riding South to North along is axis, but this would be premature, however it also contains only one magnetic spin and since we are flipping its relationship to the electron field this also causes an imbalance in the electron field along its canceling lines and makes one dominant. While one flow is always found in opposition between Electron and Proton magnetic flows from the vector model, the other is always found in attraction and not moving away from one another. So the interaction between Electron and Proton magnetism is always "repulsion and attraction". See section below. If repulsion is the inner flow then we get expansion of the field which manifests as outflow in the conscious fabric.

A Close Look at the Vector Model of Bismuth Protons Interacting Magnetically With the Iron Electrons in Rain Maker 1

In the Rain Maker 1 device the iron represents the magnetic field from the Electron layer, the Bismuth represents the magnetic field from the Proton layer, we have now separated them and can study the interactions. Remember that the inflow outflow field did not appear until both were present.

North pole of the magnets are pointing outwards on the ring. The Tempic Vector model of magnetism shows particle spins repel, and orbital spins attract. We now have a compressing magnetic orbital field setting around an expanding magnetic field from the particle level. This is the formula for outflow energy. An examination of the MEC shows a similar pattern between the cylinder and the rollers which have an over all vertical attraction, yet small opposing fields inside these.

South pole of the magnets are now pointing outwards on the ring. The Tempic Vector model shows particle fields are contracting in and the orbital field is expanding outwards. This is the formula for an inflow energy.

Since both orbital and particle magnetic flows are always interacting between Electron and Proton, we can expect the interaction to be stronger then with normal magnets having only

Electron Electron field interactions where the ones traveling directly apart do nothing.

From the above we may see that the perception and control conscious fabrics lay at 90 degrees to the magnetic vectors and operate from the scalar compression and expansion of the two magnetic fields between Electron and Proton.

When the two flows compress together we get outflow, when the two flows pull apart we get inflow.

Expanding the Tempic Vector Model of Magnetism

This is all a result of the 3 tempic vectors acting on one another in the magnetic field between Electron and Proton.

The magnetic field between Electron and Proton act directly on the conscious fabric in this way. They actually set up compression and expansion differentials in the Aether fabric which if set off balance and then looped remain fixed there.

This explains why the energy feels radiant through the entire field and not alternating along the field from inflow to outflow. The entire magnetic field of Rain Maker 1 is either outflow or inflow but not both. This field radiates off the magnetic field at 90 degrees to it in a spherical fashion and the corona gets very hot to the hands.

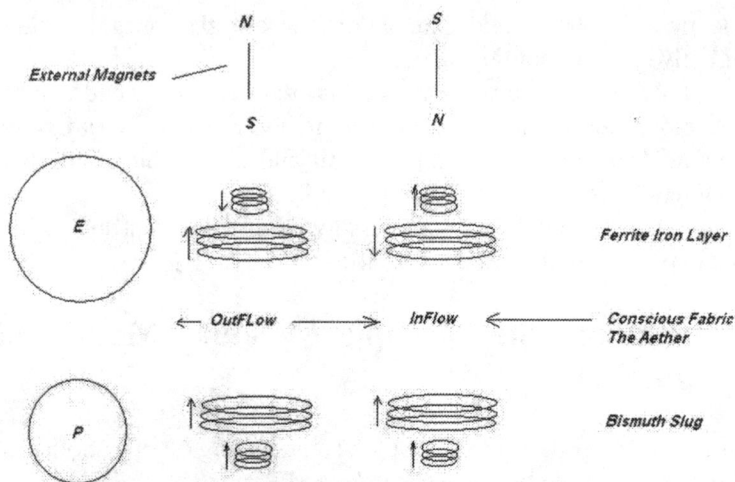

When tempic vectors move in parallel we get attraction and compression of the field. When they move towards one another we get repulsion and expansion of the field. Repulsion setting inside attraction produces outflow, and attraction setting inside repulsion produces inflow, in the Aether or the conscious fabric.

You may notice in this model the Proton magnetic layer never reverses as I thought previously. One side will always find attraction towards the electron layer.

Proton magnetism must always be turned out to interact if there is an external magnetic field present. Flipping Electrons field will reverse from inflow to outflow along the expanding and contracting tempic vectors of X an Y directions. This model comes close to explaining the nature of the Conscious fabric or the Aether and how the magnetic field interacts with it at 90 degrees through the scalar field of Proton interacting with Electron.

The Scalar Bismuth Coil

Now viewing the scalar bismuth coil from the above model we get a different view. The copper windings form layers of opposing loops up and down the coil, but at each intersection of the wires, in the area between them, magnetic fields now shoot

The Rain Maker Device

either North in or North out on each successive loop. If the fields were totally balanced then the resultant should be 0 radiated magnetic field, and yet we see one appear that flips with the polarity of the voltage present setting perfectly at 90 degrees to the canceling Electron field.

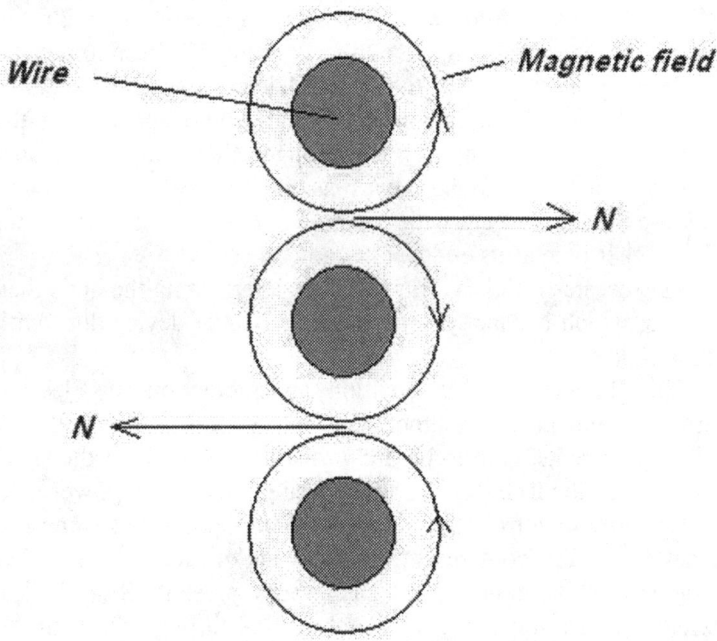

The diagram shows the cross section of one side of the bismuth coil where Electron current is flowing opposite directions through every other wire.

As the current is equal in all the wires all the magnetic fields should be the same strength and result in zero radiated field off the sides. That is the upper N should be canceled by the South right below it.

Considering that now we have the totally balanced Electron magnetic field present from the current flowing through the copper atoms Electron shells, it is this field now interacting with the Bismuth in the above fashion to produce alternating inflow or outflow layers along the coil. We discover it is enough to imbalance the Electrons field and cause one to now dominate the other. We get a total positive outflow on one polarity and an inflow on the other polarity.

While the coil appears to be vibrating the Aether or conscious fabric, it also appears to be imbalancing the magnetic field as well.

The Bismuth atoms are now seen to offer only a Proton cushion of magnetism against the reversing Electron magnetic field, exactly as we did in the Rain Maker 1 experiment. As to why it imbalanced the magnetic field it would appear that the particle spin interaction may be stronger then the orbital spin interaction, or there may be other reasons. The scalar bismuth coil radiates a 90 degree field with one pole out all around its diameter and the other pole at its ends, both ends,

This creates a field with more flux intersecting the ends then the sides, which is exactly what the Rain Maker device does with a static field.

The Bismuth coil is creating an interaction of Electron magnetism and Proton magnetism also. We could expect that if very high currents were to be drawn then we should get the same interaction as the Rain Maker device using only DC to power it.

It is now seen from the models offered that Rain Maker 1 did in fact not offer both an inflow and an outflow to see what interaction would come from them both present. Rain Maker offered a perception and control fabric link through the bismuth scalar winding, and a fixed imbalance of the standing Aether background level manifesting as Inflow or outflow. This was created by the strange shape of the magnetic field stressing the Aether, however why it did not stop when the device shut down is a very good clue as to how consciousness works within its fabric. It may have simply been that I was feeding it with conscious form and as I continued to work with it, it remained stable. This effect was previously noted in using the scalar coils in the mind. A strong inflow would stay present in the mind as long as focus was directed towards it. The other explanation is that the Scalar magnetic flows were in fact still flowing but self canceling perfectly so could not be observed, this is hard to believe because the magnets were no longer present, but may also be true.

More on the Scalar Bismuth Coil

Due to more experiments I have noticed that it is the Scalar wound coil on the Bismuth slug that is causing the effects. A function generator is not even necessary for this to manifest, simply short the scalar coil, place it inside the iron cylinder with magnets pointing inwards and the energy begins to heat up very quickly. The application of high DC currents on the coil does not increase the effect with either polarity. The electron generated currents do not aid the process.

It would appear that the Bismuth is the generator of the important magnetic field.

The interaction is happening between the Bismuths diamagnetic field and the scalar coil.

This would lead to the possible belief that it is the bismuth Proton layer directly responsible for altering the Aether, gravity and time flow rate when acted on by the outside magnetic field The Electron shell is providing the EM field and the Proton shell the Aether[Gravity Time] interaction.

Something inside the nuclear strong force area is coupled to the Aether, it is only altered during an interaction with the diamagnetic layer.

Bismuth [slug]
Magnetic moment Proton - [+ 4.54440]
Nuclear Abundance - [%100]
Spin - [9/2]
Atomic weight 209
8.356 MHz

Aluminum [tube]
Magnetic Moment Proton - [+ 4.30869]
Nuclear Abundance - [%100]
Spin - [5/2]
Atomic weight 27
13.550 MHz

Copper [scalar coil]
Magnetic Moment Proton - [+ 2.87549]
Nuclear Abundance - [% 69]
Spin - [3/2]
Atomic weight 63
13.788 MHz

If NMR is a factor then it is totally internal to the system as no frequency injection is necessary.

The greatest difference in the Aluminum and Bismuth is its atomic weight or mass, as the magnetic moments are almost identical. Also the spin is higher for Bismuth.

When you energize a coil with Electron magnetism you get a counter EMF produced that acts to resist the incident signal.

This is Electron action acting against coppers inductive proton layer creating an opposing effect.

Induction Effect

We are familiar with the nature of induction with respect to Electron flow.

My version of induction, Electrons begin to flow through a medium like Coppers Electron shell. The flow is reflected back from the Proton layer as a countering force slowing the change of current. Inductance comes not from the iron core but from the magnetic Proton layer of the Copper itself. We see air coils offer inductance as well, and it is assumed that a coil in empty space would also offer inductance. Inductance is a function of the Copper itself operating between Electron and Proton magnetic forces which counter or attenuate one another lowering the energy in a circuit [back EMF].

Now in the Bismuth coil, we have set up a Proton interaction with another Proton layer in the Copper and a Copper Electron canceling layer. Any energy moving between the Bismuth Proton layer and the Copper Proton layer may pass through the Coppers Electron layer but will be canceled or reflected back in to the Coppers proton layer.

The Rain Maker Device

Bismuth is unique in that it offers a diamagnetic Proton magnetism that can be accessed outside or along the surface of its material base. It is this field in motion of its own nature that is acting in the scalar coil and being reflected back into the Coppers Proton layer that is powering the Vortex generator.

The scalar coil is the link between Bismuth and Copper Proton magnetism. The regenerative building effect is happening between the Bismuth and Copper wire that has its Electron field canceling. This is the true source of the interaction that is causing an energy envelope to pop out of the whole unit at approximately 3' diameter.

The surprise is that in the interaction of two inductive elements, pitting inductance against inductance does not create a degenerative effect but an over unity effect. The coil is self powered as long as the scalar winding is shorted. Opening the scalar coil halts the buildup of the field by allowing the Coppers Electron layer to form countering voltages that sink the energy.

Now with our coil we are hitting it with a diamagnetic field [Bismuth], and getting a counter field back at the Proton layer. So the Copper wire on the Bismuth core is a Proton - Proton magnetic interaction. This may be the effect we are seeing here.

Although a canceling coil tends to reduce Electron magnetism it seems to be doing something very different with Proton magnetism. It is actually forming an Electron energy mirror between the Proton layers of the Bismuth and the Copper where energy can flow between.

Placing the "South inwards" field along the coil is aligning all the Protons in the Copper and the Bismuth into the correct alignment for the Aether pumping diamagnetic effect which are jumping through the copper Electron shells.

The Vortex Generator

The Vortex generator mimics the Scalar Bismuth Coil magnetically with its unique shape, only now it introduces the Iron Electron field into the curved magnetic field. A magnetic donut with one pole outwards and the other pole inwards to the donut hole.

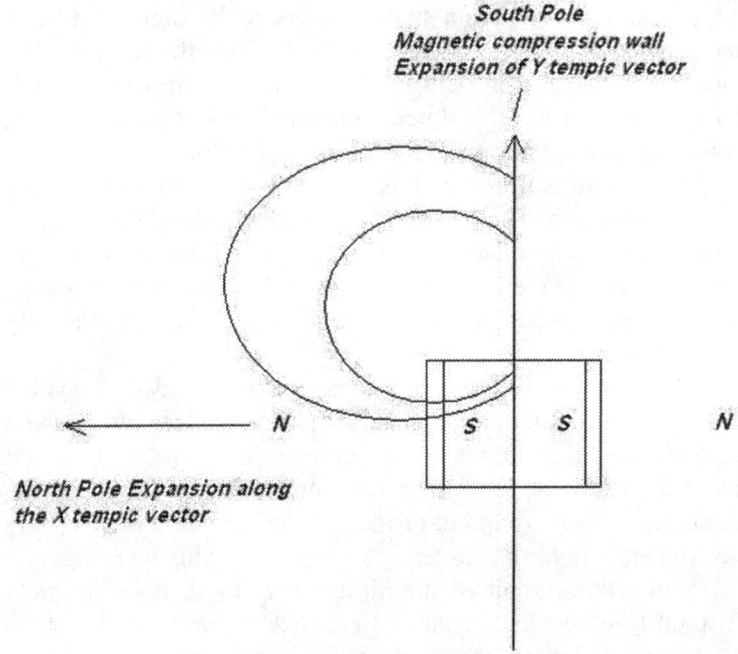

Tripole Field created by the Cylinder

We see the magnetic field that is set up in both the scalar coil and the Vortex generator curved such that one pole is compressed in wards and the other pole is free to expand outwards. The field in the iron cylinder has become its own tripole by compressing like poles inwards all the way around. This creates an expanded field outside the unit and an expanding pole sticking up and down from the cylinders ends. These two expanding tempic vectors are setting at 90 degrees to one another, exactly as in a tripole magnet configuration only in a complete circle. The inner field is receiving a compression in X Z vector motions and an expansion in the Y vector. The field outside is experiencing an expansion of the X Z vectors and a release of the Y vector.

The two poles of the magnet are having exactly opposite tempic effects in the 3 directions of the magnetic field.

The Bismuth sets in the first field compressing along two tempic vectors and expanding along one, the Iron sits crossing both because it is magnetic but mainly resides in the other with a reversed tempic vector result.

It is noted that this magnetic field shape alone may help explain a time effect for energy spinning in the X Y plane of motion across a magnetic field pointing inwards. However add to this the Bismuth Protons are setting in the strong force area with a lowered gravity and we see more. Torsion is coupled through the magnetic field, there will now be a flow of torsion energy between Irons Electron field and Bismuths Proton field. Which of these two actually cause the CU to begin to form is up in the air at present, until more experiments are done with other shaped devices. At least we have identified these two parameters for now.

The observed interaction that it was possible to generate an inflow with far less magnets then it took to generate an outflow indicates that we may want to reverse the materials in order to generate an outflow easier. Both the effects are very probably involved in the process. Thus placing the Iron in the compressing field and the Bismuth in the non compressing field may be a better choice for an outflow field generator. We also find in the Searl cylinder the copper is setting outside the iron layer.

Power Coupling

Since most scientists want some measure of proof, although even these days lighting a large light bulb on video is no longer considered solid proof, I feel if we could couple some kind of power off the device while a Vortex is opened it would be a good solid proof of concept only if accompanied by a solid theory that shows we can alter parameters. This will give anyone the ability to design a device.

Receiving scalar energy, a pickup and loading device, is essential for a first step as well as a way to sense the presence of the torsion energy for those who are not able to actually feel it with their hands.

Even if we let the CU fully form and it becomes strong enough to power a craft we would still want a way to channel off some energy for use in electronic gear or other things. The main problem is that the free standing CU does not even deflect a compass, so how can we couple to it?

We discovered that the interaction between Proton and Electron magnetism is the key to generating such a field so it may stand to reason that this would be a way to couple energy from the field as well after it is set up.

If we want to couple motion then we have to set up a CU with a spin.

If we want electric power then we need to somehow reclaim both opposite fields from the magnetic canceling vibrations.

Tungsten is said to alter it's resistance in a torsion field, and this could work for a sensor but this is a slow operation and takes time for it to alter its resistance.

Sweet left us a clue with his bifillar setup. Another method I may suggest, use iron wire next to copper wire in perfectly balanced coils like a Stubblefield cell.

We know that each one will couple at a different level to the scalar field and between them we may produce an effect particularly if we set up a magnetic field between them. A coil with one side copper and one side iron, may give a more immediate result.

Setting up two CU's that spin opposite directions may produce magnetic pulses. The 3SD. Then Power can be tapped like Searl did with pickup coils.

Bearden indicates that at the interference point between two scalar fields a magnetic field may form.

Also we recognize that the inductive metals [Aluminum Copper Bismuth] are Torsion coupling materials when a magnetic field is present. This can be used to produce motion, but also electric current if the Copper is held stationary in a changing magnetic field. Problem with this is that the scalar fields are canceling along the voltage plane and normally no imbalance is present. Directing the scalar field along a surface with bismuth and aluminum coming together at 90 degrees to the scalar field should produce an imbalance between them, however the result will be atomic spin interactions and not electric.

We are finally left with crystals, and the crystal sphere, which is in fact shown to amplify the torsion field, or scalar waves. This can be sensed in the body as an indicator of a field that is building, however it too is very probably scalar in nature. Tests must be done with spheres to see if voltages or currents are emerging in the vibration of crystals with scalar waves.

The wave formed in the conscious fabric would appear to be a 2 dimensional compression along the two tempic vectors setting at 90 degrees to the magnetic B field between electron and proton. This is very similar to the model of gravity which sets at the precession area of the electron shell and modulating this area with a small circular movement along its center. This being the case gravity may actually be the interaction of the Electron and Proton spin magnetic fields interacting at both the particle layer and the orbital layer compressing against one another. In a normal non magnetic atom the torsion between electron and proton is constant. The two fields would produce a constant Aether interaction leading to the "Zero point field" in normal matter. This would represent a constant scalar pressure between the two forces of particle and orbital fields pressing on one another. Protons and Electron normally couple in a dance of spinning dipoles. Now we know that Proton always turns to face Electron in this dance, and in the process a conscious Aether pressure or time flow rate is established setting at 90 degrees to the line between them. Now it is suggested that we may be able to think of this as all one field. The torsion gravity field or the background Aether Zero Point. Gravity time alterations are a case where the Zero point field is changing slightly to a Non Zero Point state.

Whether the magnetic fields are vibrating the Aether or the Aether is vibrating the magnetic fields is not clear, however it would now seem that if we place matter in an alternate time flow rate, it's atoms should all begin to come up to the same speed, and this is density sliding. By introducing a stronger magnetic field then normal we cause a stronger coupling between them, by introducing a strong scalar field we allow energy to "slide" through the Aether between them.

Suggesting a Math Model for the Zero Point, Gravity, and Scalar Canceling Voltage

According to Wilbert Smith the tempic vector and the voltage vector are the first two forces of nature. But now we see how the magnetic fields compress or expands these two forces, not only a precession, but as scalar canceling force along a plane at 90 degrees to the two magnetic fields involved, or longitudinal compressions and decompressions between the flux cylinders of the two magnetic fields.

This is the scalar interaction of magnetic fields along the [voltage / gravity] plane of motion:

Longitudinal compressions and decompressions between the flux cylinders of the two magnetic fields operating between Proton and Electron creates gravity and altered time flow rate in the Aether fabric.

Gravity always manifests in a spherical field.

Electricity also manifests in a spherical field.

I now believe I have discovered the difference.

Gravity and time, interlocked forces actually set in the fabric of the Aether, or the conscious layer, as scalar or opposing pressures of the magnetic field at 90 degrees to the B field which is setting in the physical fabric. Gravity is related to the square of time flow rate.

This means that space is permeated with a background scalar pressure which can be thought of as the "Zero point of space" or ZP. If TFR [time flow rate] = C [light speed] then Gravity = $\sim$ C $^\wedge$2. Or Gravity = $\sim$ TFR $^\wedge$2. $\sim$ = some function unknown as of yet.

My first original equations!

However we know that time flow rate should decrease as gravity increases. The inverse function. Since gravity is two dimensional laying across in the Aether, to its Electric counterpart in the physical, and the tempic vector is one dimensional. Gravity is then related by a not squared function to time flow rate, but an inverse squared function.

The Rain Maker Device

TFR = Time Flow Rate
Gravity = 1 / TFR^2

It would appear that matters electric scalar pressure on the time flow rate normally increases gravity. The TFR [time flow rate] is less then light speed for spin at the Electron and Proton layers. We interact with the time flow rate in 3 dimensions and thus through the magnetic field, however gravity interacts through only two of the dimensions. Therefore relative to experiencing the time flow rate of the sensation of time, it is linked not to just gravity but also the B field. It is the B field or the third dimension that sits crossing the density barrier, on each side of the B field from Proton to Electron there is a different TRF and Gravity, but the magnetic field can link across both. From our perspective of clocks then. (cube root [Time flow rate clock speed]) ^2 = 1 / Gravity

This accounts for all three tempic vectors of the magnetic field and the actual 3D effects of the tempic vectors as they manifest magnetic fields.

The math model is still severely short because we have not allowed for a gravity moving the opposite direction from Zero Point energy level and creating antigravity because in our equation Gravity must always be positive because it is a result of a squared function of time. Thus the Zero point of Space must equal some positive value and a negative gravity is really simply a value lower then space, and the true model.

Zero Point - Gravity = 1 / Time Flow Rate^2
ZP - G = 1 / TFR^2
[The Lowrance Equation]

Now we see gravity is negative and thus sucking below Zero Point, which becomes variable at the different densities. Gravity is seen as an alternate ZP zero point from free space in a region of matter, or now a Non Zero Point. With this math model we can have a positive gravity or antigravity based on the inverse square of the time flow rate, which has been observed. There may also be a constant that needs to be applied to unify the units of measurement, however this equation comes close to my feel for the link between gravity, zero point energy and time flow rate. ZP is like a background voltage would be in the physical sense. ZP is expressed here as a squared function of the Aether tempic vectors rather then a cubed function, just as voltage is a squared function in the physical fabric.

Zero Point Math

We now see Electron setting at "ZP - Electron gravity" and Proton setting at "ZP - Proton gravity".
Both are connected to the 3rd density Zero Point the same:

$ZP = [1 / TFR (E) ^2] + Gravity (E)$
$ZP = [1 / TRF (P) ^2] + Gravity (P)$

And this may be the first constant for a Density:

$(1 / TRF ^2) + Gravity =$ (The Lowrance Constant) LOL!

The Zero point Aether background energy level, related to gravity and time flow rate as a squared tempic vector value.

$(1 / [Time Flow Rate] ^2) + Gravity = ZP3$ (The Lowrance Constant) or [Zero Point for density 3]

Now I suppose I will have to plug in the numbers and come up with actual value for the Lowrance constant to become official, based on the loss of mass inside the nuclear strong force area.

Scalar Magnetic Math

If in fact the TFR, gravity, and the ZP fields actually lie in the fabric of the Aether then a function of phi may be applied to relate them to the magnetic fields effecting them on this side in the physical, the electric and magnetic forces. That is the Electric field in volts applied to a scalar setup where it cancels would

The Rain Maker Device

result in a (1.618 / 1) ^2 ratio to the gravity component it produces, or the change in ZP the zero point field. We have already seen this in Andrews 3 6 relationship I believe. An electric shift of X volts could generate a gravity shift of X (1.618/1)^2 in a scalar interaction between Electron and Proton.

X volts[physical fabric scalar canceling] / (1.618)^2 = gravity shift of X [Aether fabric] * some constant
[The Bellon Equation]

Scalar magnetic math, what a concept!
It is all there in the models, if proved true the math will come that fits the models. We do not need the math to ever limit our observed models, we only need adjust the math to fit the models. It is not the math that governs the models, the math describes the models which exist independently of the math. An observed anomaly means the math model is wrong.

Kosol and Koeun Ouch, Vince Panella and David Lowrance

Measuring the Background Aether Pressure, Density Factor

Any scalar canceling coil using an inductive metal as a core set 90 degrees to a normal iron core coil could be used to measure the background Aether level. From this can be determined the density, the gravity, and the time flow rates for where the coil is located. The 90 degree field setting off the side of the Bismuth scalar coil is the product of the Electron and Proton interaction, or imbalance. We could also just weigh the coil, however a ship in space could not do this. The measurement needs to be self contained.

We would expect that as we slide along the "Density scalar" towards a higher density, the background Aether pressure would be altered. Gravity scalar decreases, time flow rate increases, and space becomes more dense, as matter physically shrinks and we become more aware. This would affect all our instruments but what would become altered is the gravity and time flow rate values. We must somehow relate this to the magnetic field as measuring the gravity between two small objects is not practical, nor is designing a device that can sense an altered time flow rate because it is affecting us as well.

We must use a device that has both Electron and Proton magnetism interacting off one another and find the parameter that is altered by the Aether background scalar. The only obvious method may be the scalar transmission and reception moving through this medium. Planks constant would be altered. Although frequency would be raised, we could not perceive this due to an increased time flow rate. The interaction that would be altered is Electric to Magnetic force interactions due to the altered tempic vectors. We may expect that a smaller voltage would produce a larger magnetic field and this is the only easily monitored relationship. The B field will be stronger as gravity is lowered.

This model assumes that as the overall time flow rate increases along with the magnetic field strength, the voltage and gravity scalars will drop together. This seems a fair assumption based on their parallel relationships.

We set up an AC signal into a scalar bismuth coil, this will measure both Proton and Electron fields together, then we set a normal iron core pickup coil at 90 degrees to it, and the output signal will vary as density is altered. The magnetic field operating between the two coils will be increased or decreased, although the voltages will appear to be effected the same, the coupling between the coils will be altered due to the increased magnetic field. This is the H coil that appears in Smiths work, only the core must have a bismuth layer and not just iron ones.

The H Coil

The H coil would have a center piece consisting of a small iron pipe coated with a bismuth layer. This is wrapped with a Smith coil. The outer H structure is iron, wrapped all the way around with a normal coil winding, or four pickup coils in aiding wrap on each H leg. The two coils set at 90 degrees to one another. The 90 degree flux will travel easily through the iron inside the scalar coil and appear in the H structure iron core with poles to each end. As AC energy is placed across the Smith coil at the center an output is seen on the normal wound iron coil indicating the Proton to Electron 90 degree field moving perpendicular to it through the whole structure. As density is altered the magnetic coupling between the coils is increased and the output voltage is seen to rise above the normal voltage ratio moving through the coil before.

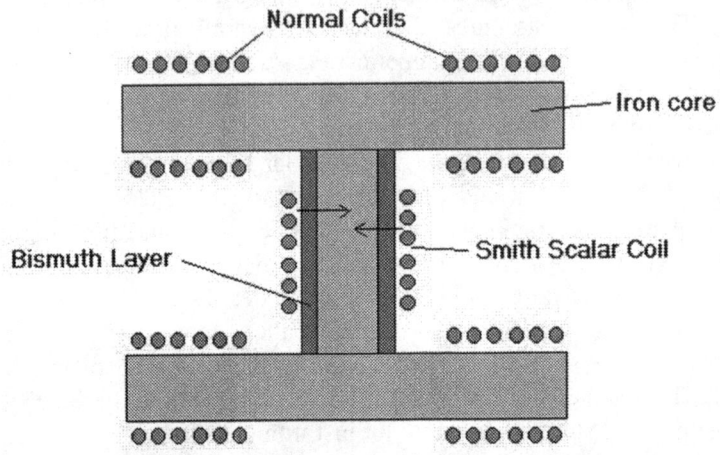

H coil for measuring Density or Aether background Pressure

The Scalar coil produces the 90 degree field against the Bismuth layer, which then passes through the iron to the whole coil where many turns may bring the voltage up.

This coil structure would also be seen to harness the 90 degree flux for amplification and clear analysis. It should become the basic instrument for exploring the density "sliding" devices. It's input to output ratio will be a tempic squared function [voltage] and may indicate the inverse of "gravity" directly. The square root will indicate tempic flow, and this number cubed will represent the magnetic field strength as well as the overall 3 dimensional time flow rate experienced within that magnetic field.

A device like this should be constructed for testing of the vortex generator.

Protonics

In my first paper and study on the forces of nature present in the universe, I touched on this principle of cold electric energy or Protonics. It would seem that now we stand on the brink of discovering this new world of energy transfer via the Protons shell. Proton energy is easily tapped using the high nuclear magnetic materials, Copper, Aluminum, and Bismuth. I believe that the field of Protonics will be the greatest discovery of the 21st century and completely change the way we think of and interact with the world.

Electronics - The study and mastery of the nature of the Electron shell, using it to move energy from point to point mainly with free Electrons and radiation of photons in the "medium" formerly referred to as "free space" or the "Either" and the magnetic field. The equations are abundant for Electronics and magnetism as related to "c" or the present model of light speed.

Protonics - The study and mastery of the nature of the Proton shell and how energy is moved through it as it sets in the strong force area of the atom, and propagates energy through the "scalar background medium" that lies under or behind "free space" in a higher density with a faster time flow rate using the diamagnetic field. The equations for Protonics are totally absent to the engineer at present and the scalar devices are little understood as well as the speed of propagation being 1.5708 to 1.618 higher then "c".

Diamagnetics - Study of the nature of Proton diamagnetism which always provides a repelling force to an Electron magnetic field. Diamagnetism is linked to motion [torsion or spin] and is not a constant force but is altered by motion. This has been missed by most classic explanations of diamagnetism which claim diamagnetism is the result of Electron interactions and is a very weak force. Diamagnetism is seen to shoot pieces of Aluminum at "bullet or projectile speeds" it is only a weak force in a stationary magnetic field but becomes a very strong force in the presence of a moving magnetic field. Equations to describe this hidden force relating torsion to diamagnetic force in various materials are presently missing from our primitive technology.

After my own research into diamagnetism I am truly appalled by the classic definitions of this force which may hold one of the keys to over unity. In the materials Copper, Aluminum, and Bismuth consulting an NMR chart will quickly reveal the nature of diamagnetics is the high magnetic moment of the Proton field and has little if anything to do with Electrons which always seek attraction to external Electron magnetic fields.

As the Proton sits in the strong force area of the atom where mass is reduced, and thus gravity is lower, we will have to account for more of the forces as we begin to understand manipulating its fields. Tom Bearden warns of the dangerous nature of the "conjugate waves," supposedly moving backwards through time. Tesla however did not believe in time as a dimension and strictly held that all is motion in the Ether or the background "Medium". Wilbert Smith seems to agree with Tesla on this and describes the world as based solely on "spin" and altered "time flow rates." Everything moves through time, but not at the same rate. Gravity lies in the difference. I am now forming a model with two levels of Aether that lie is slightly different densities, and therefore the energy propagating in each one does not interact.

On Entrophy and Awareness

"Entrophy is a state of negative mind, an illusion." All is motion, and motion can not be stopped at the deepest levels of matter." This is a true Law of nature. "Nothing comes to rest, ever". "All is change, and change is constant." There is no such thing as an "isolated system," all systems composed of atomic particles are

The Rain Maker Device

connected to forces we can not at present understand, all atomic particles move through transitions of disappearing and reappearing in this realm. Consciousness must slide deeper to realize these truths. Awareness is not fixed so as to be able to determine exact truth for all. A scientific model needs to accept this parameter. This is opened ended science, one that accepts its present understanding is not a "law" but is fluid, just as reality is fluid to the level of awareness achieved." "Experiments involving Awareness or Consciousness do not produce consistent results, yet they should not be dismissed for this reason alone, but awareness must itself become a variable of the experiment."

Myths

The myth of "reverse time flow waves" in the Proton or Electron shells, and the myth of the "scalar field" will be hard to dispel. Observing the nature of the "scalar waves" in the Electron shell and how they interact with the Proton layer one comes to a slow realization that the energy of a scalar canceling winding on a coil does not actually cancel the energy present in the magnetic field but simply pushes it through to the deeper Aether medium of the Proton layer.

It is through the scalar canceling magnetic field that the energy is transferred from the Electron layer to the Proton layer. Energy is not lost or created, it is transferred between these two atomic layers through the magnetic field that couples them together in the never ending dance of the atom.

We may get confused by this transfer and make up alternate models of "reverse time flow waves", "scalar fields", Electron scalar canceling fields, or potential scalars without motion or vector, to try and explain the observed phenomena, however sooner or later the truth will be realized. I believe that there is really no such thing as a "scalar field" in the most exact interpretation of the term, without a "vector" and a "scalar" there is really nothing at all, no force. I also believe that all the forces holding this reality together travel through time with us, not backwards, and thus no force is lost. Wilbert Smith gave us a model of tempic vector, voltage vector, and magnetic vector, with "spin" and only "spin" being the one thing all awareness can agree upon. There is no such thing present in the universe as a potential having no motion, because at the roots of all matter at the smallest places of existence all is spin, and all is in motion and this is the root of time flow. This is nature. The voltage vector, the second dimension, cannot exist without the tempic vector, the first dimension it is built upon.

The energy from a scalar canceling magnetic field does in fact go somewhere real. Because we have not found an adequate name yet for the medium of energy transfer that seems to bypass the conventional EM path through space we resort to models of two static forces propagating through the same space but do not interact with other EM fields.

The True Nature of the "Scalar Canceling Field" and Density

Study of the scalar canceling coil tells us that the Electron magnetic field can totally disappear from view in the normal sense of power transfer that we are used to monitoring and then pop up at a remote location set up to receive this energy. It seems to bypass the normal Electric current, voltage, and RF energy which can be intercepted with a Faraday cage and has been called [tunneling]. We felt that there must be a hidden medium and started to develop models to try and explain this just like the first Radio engineers. The presence of a mysterious scalar canceling field, which cannot be measured, yet is present because we can transmit and receive it across space and even through objects, as though the energy were somehow leaving our known universe and reemerging around it. Tesla observed this as a stinging on his skin with high voltage surges, the energy which could jump through glass shields when the switch was thrown and he recognized it was not normal electricity.

On Tesla

Tesla indicated it was the Ether, the "Medium" through which these disappearing forces traveled. Tesla clocked the transfer "medium" speed for the "stationary ground wave" at 471,240 Km /sec and the speed for the Electron transfer medium at 300,000 Km / sec. [Tesla patent 787,412]

Although the patent indicates he did not believe this to be a space propagating wave but moving through the earth directly, at faster then light speed. He presents this wave as being like a standing wave found on an antenna, only moving directly through the earth and not in the air or free space. He later describes this as a spiraling energy from a center zero point, and his coil designs in this patent were apparently based on this model. His descriptions have been compared to a "longitudinal" wave and now many words have been added to our alternate energy vocabulary. Tesla wave, longitudinal wave, scalar wave, cold electric wave, some are not proving to be particularly accurate, however the terms all point to the same principle. A wave generated by the Proton medium traveling faster then light through an alternate medium of transfer where there is no degradation of the signal strength across space.

It should become apparent the Proton energies travel through a different medium then the Electron energies. One that lies in a different density where relative to us time flow rate is higher. Tesla left us the actual speed differential. Tesla's longitudinal waves left a charge in any metal objects for a great distance around the field generator, it was not balanced but produced an offset voltage because it did not equalize the Electron and Proton energies perfectly as a true scalar canceling potential. His own description of this was a "disrupter." He utilized the fast "impulse" of only the Electron energy to create the effect.

Tesla also made some of the same observations about this form of energy "Protonics" as I have rediscovered. The sensation of "physiological heat", "pressure" or torsional forces and at his high levels of energy even pain and the vibration of objects. At higher frequencies the presence of white light filling the room.

Connecting the Models

I have recognized that there are really two "Mediums" present in the operation of the atom and not merely one. Energy can move between them, but it does not "cancel." The scalar coil is a transfer mechanism for bouncing energy between the two "Mediums" or alternate "Densities". The true vehicle is the magnetic field which is the only field that can span both "Mediums". The B field as it sets between the Electron and the Proton that share a B vector and illustrated in the Time Vector Model of Magnetism. Actually there are two tubes the B field uses to connect between Electron and Proton, one is a result of particle rotation and the other is the result of orbital motions. It is the interaction between these two tubes of magnetism that transfer energy into or out of the so called scalar transfer medium, setting in an alternate density. Scalar waves are actually longitudinal pressure waves of some sort, probably with both scalar and vector, in the Aether medium that lies at a deeper level then the space transfer medium of the Electrons EM fields formerly called the "Either". The time flow rate differential is $471,240 / 300,000 = 1.5708$ [Density Proton / Density Electron] Time flow Rate [Aether medium ratio]. This is slightly different then the predicted 1.61803399 from the phi density model, however the difference [0.04723399] is less then 5%.

EM travels through the spatial medium colliding with anything in its path and forming interference patterns. The background medium is the deeper Aether, and Protonics is the method of propagation not Electronics, or at the least an interaction of the two. The Protons energy transfers are in the Aether and seem to move around or behind the spatial fields and the normal Electron channels of what we now call "free space" and at an increased velocity seemingly impossible from our "Einstein point of view" about light speed being the constant of the universe "c." Tesla already showed us the true nature of "Density" and "Time Flow Rate."

The forces connected with the Electron spatial fields are Radio RF, Voltage, and Magnetism, while the forces connected with the Aether of the Protons world appear to be similar, in the "interaction" we find Gravity, Time Flow Rate, added to the list, and all the other "strong force" energies we may discover along the way such as a time sheer spherical field, or the basic UFO field, which is modeled as an enlargement of the strong force area of the atom itself.

The properties of the Proton "medium" are not clearly understood at present, however they will be as Protonics develops.

The two particles, Electron and Proton, inside the atom sit in different "mediums" or Aethers in slightly different densities. The Aethers do not interact directly, and Electron signals can move right past Time Flow Rate vibrations ignoring one another, just as Gravity can move right through a magnetic field and not affect the magnetic field in the least. There is no interference pattern resulting in the crossing of these very different kinds of fields.

Our physical universe seems to lie in the crossing of two Densities or two realities of substance, the Electron is grounded in one and the Proton is grounded in the other. They are coupled through a magnetic field that spans both. On one end of the field, Electron, we find our normal background Zero Point, with its relative Gravity and time flow rate, and on the other side we find a slightly lower Gravity and a higher time flow rate grounded in another Zero Point and forces in operation that bypass the outer medium.

Energy can be transferred through either "Medium" as has been now shown with the scalar coil, and it can be moved between them, but it is not canceled or lost. The human does have means to experience both mediums and make a clear identification.

Regenerative Affects of the Proton Medium

The Protons Aether medium, or more dense space, is not the same as our normal EM space. In the Vortex generator we find two Protonic fields interacting off an Electron scalar canceling coil like a mirror through the coupling magnetic fields. The Bismuth's diamagnetic energy is bouncing off the Magnetic canceling Electron layer of the Copper coils windings and back into the

Coppers normally inductive Proton layer and also into the Aluminum Proton layer. The two inductive elements do not fight or lower one another as with normal inductance of the Electron interaction with it. The Protons two motions of magnetism are not opposite like the Electron but both spin the same direction and both materials have a different "spin" property. There is an increase or slow building of a 3' field that appears in the interaction, but this field seems to lie in the Protons Aether and not in the Electrons Aether. It does not deflect a compass or interact electrically. This does not mean it is a "scalar field", the field has real forces coupled to it, and they do not cancel. They merely do not interact with normal Electron EM fields, as Gravity does not interact with a magnetic field. The field produced in the Proton Aether of the Rain Maker vortex generator is "balanced" and does not create voltages in all nearby objects like Tesla's disrupter field generator, but it does produce the perceived heat sensation very intensely and very constantly, not as "impulse" events when a switch is thrown, but a constant spherical shaped field. It also interacts with "intention" or the will of the person setting within it.

Another observation stated by many who play with the Scalar cancelling waves, or what is now referred to as the Longitudinal wave "T wave" [Tesla wave], is that there is no loss when they propagate due to heating of the medium as with EM that drops off as it travels. This kind of wave is then well suited for a "no loss" type of oscillation.

Also we have the examples of Searl, Hamel, and Floyd Sweet to indicate a regenerative effect is possible in the inductive metals without the transmission of EM fields into nearby metal objects. I believe that correctly identifying and removing the mysterious descriptions of these interactions is the beginning to understanding them and gaining a useful model.

Clearly identifying that the two transfer "mediums" are not the same, is a first step. Beginning to understand the deeper one will bring the mastery of Protonics, Gravity and altered time flow rates, and possibly even density travel or UFO propulsion systems.

Diamagnetic Regeneration

Copper [wire], 3/2 Proton spin, magnetic moment 2.87549
Aluminum, 5/2 Proton spin, magnetic moment 4.30869
Bismuth, 9/2 Proton spin, magnetic moment 4.54440

    The idea of diamagnetic regeneration is using a scalar coil to cancel the Electron interactive magnetic field, which then acts as a mirror to bounce energy between two diamagnetic materials with different spin qualities to form a building field in the Proton Aether. If the energy crossing becomes strong enough then one layer of the scalar coil can be opened to tap the Electron Flows by isolating them with diodes and reclaiming both flow directions.

    Due to the short field reach of the diamagnetic field [Proton field] layering must be used to increase the 90 degree field offering the wire a close proximity to the diamagnetic material. The scalar canceling element is a multi layered coil using layers of the dissimilar materials alternating with copper wire layers which are scalar wound and shorted at each layer for maximum Electric field cancellation in each layer offering a very low resistance.

    The energy transfer is controlled along the 90 degree field which pops out of the active scalar coils. Either strong magnets can be used or a ring of iron core electromagnets to set up the flow direction field for inflow or outflow, or a combination of both along the outer ferrite ring. It is hopped that as the Proton field builds, oppositely moving currents in one of the scalar coils will be at a frequency coherent enough to tap with diodes such that only this layer of the coil will no longer cancel the Electron magnetic field. Determining the correct layering for each material is the challenge as well as where to introduce the iron layers which have been shown to be necessary for the flow to begin.

    Layering presents construction challenges but the increased diamagnetic effect should be well worth the efforts as this effect will be directly related to the area covered in the small distance where the diamagnetic field is located very near the active materials used.

Observed Phenomena

The first observation I made as to the Proton Aether medium are that a stationary field may form as a spherical shape standing in space with the same motion as when the field was created. Since the field has form and can be felt as building over time due to a Protonic interaction between two non powered metals, I observe it is what we would call an over unity effect. I do not believe that it is motionless or has force without a vector, this would seem unnatural. However this may be the nature of the Proton Aether interacting with a reflective Electron Aether through the magnetic field. Since it is setting at a higher density on the Protons end whatever is happening is causing a change of the Protons Aether field to form a spherical energy that is not the normal background level. Whether this is altering the Protons Aether or the Electrons Aether, or the Control or Perception fabrics is not truly known. Until we see a gravity or time effect our opinions are premature. A gravity effect would indicate a Proton Aether field, a voltage effect would indicate an Electron Aether field, a magnetic field effect would indicate possibly one or both. As we have never seen this sphere shape appear in anything but an Electric or a Gravity field, I have high hopes that Gravity will be the potential we are affecting.

The Protonic Forces

It is hoped that with time all the Protonic forces will be clearly identified and this list will grow with an understanding of how the forces interact. However for now it is important to realize that when dealing with Proton manipulation one must consider more then just the Electric or Positive potential of the Proton. Wilbert Smith identified 12 dimensions, [see references below] or layers of forces in four fabric groups of three each, all in quadrature alignments. It is suggested that between the two forces of Proton and Electron we may discover all 12 at some point. As we have covered EM, as well as an introduction the Perception and Control fabrics, next are the not so well understood forces and how they interact, the tempic vector and spin.

Torsion Spin

Protonics converts torsion to a diamagnetic field

Torsion leads the diamagnetic field in real time

Increase of diamagnetic force is a direct function of the motion placed on the magnetic field

    Torsion or spin of momentum is critical to the Protons diamagnetic coupling because the largest mass of the atom is contained in the nucleus. As we alter the angular direction that an external magnetic field crosses a diamagnetic nucleus we get a delayed reaction. The magnets dropping down the Aluminum tube show this delay. In Electronics it is noted that in an inductor or transformer the current leads the voltage by 90 degrees of the frequency applied.

    This delay seems built into inductors and we have identified that it may be coming from the mass of the Proton having to physically turn into alignment with the external magnetic field moving through it. As it turns it generates an always opposing diamagnetic field from the Coppers Proton layer.

    Further in experiments with spinning diamagnetic materials like Copper we observe another very important interaction. The Proton diamagnetic layer is doing a conversion of forces with a time delay built in. If we bring a magnet close to a stationary copper cylinder we observe no forces worth noting. As we spin up the Copper cylinder we observe an increasing diamagnetic field that seems to build with the RPM. It always opposes any magnet in any arrangement of the poles.

    That is, Protons magnetic fields are diamagnetic but the strength of the field is coupled to motion. The important observation is that the diamagnetic field lags the torsional motion that caused its conversion. As we add more spin velocity to the cylinder the diamagnetic force increases. This process requires the nucleus of the atom to alter its spin direction and causes a delay of the resulting diamagnetic field build up. With spinning Copper cylinders the repelling force will extend many inches out from the Copper.

Within this interaction of motion and magnetic fields we find a method of increasing the diamagnetic force available for interaction.

Searl Disc Revisited

Another instance of diamagnetic regeneration.

In the Searl disc, once motion is present, it is this diamagnetic lag or torsion leading the diamagnetic field conversion that may be causing the parting side of the rollers and cylinder to push away harder then the other side is resisting moving together. As the magnets and Copper come together, the copper atoms must turn 1/4 turn down to align with the opposing magnets, they begin to couple and convert the torsion of spin into a stronger diamagnetic field.

On the closing side of the rollers the Copper Protons are experiencing a torsion, on the opening side of the rollers it has been converted to diamagnetic repulsion which causes a stronger magnetic push along that side. What ever the delay is, sets the diamagnetic forces slightly off centered between the roller and cylinder sustaining motion.

This would indicate another regenerative use of diamagnetism, as the Protons field is not depleted or used up in the process.

The Copper layers of the roller and cylinder are two diamagnetic layers interacting with opposing magnets. This conversion of torsion to a stronger diamagnetic field with a built in lag time may be the key to the Searl disc.

Generating a Strong Diamagnetic Field

Getting the diamagnetic field to emerge outwards from the atom will offer a means to study it closer.

The Searl disc gives a clue for a device to study the diamagnetic properties of the Proton in Copper, Aluminum, or Bismuth. Putting torsion on the magnetic field cutting through a diamagnetic material will cause the field to expand outwards with great force. It is hoped that the conversion will also contain a time lag we can use to our advantage. There are two methods immediately available. Spin magnets past the diamagnetic surface, or set up a moving magnetic field in a coil. The Searl disc uses only two fields at 90 degrees to one another, so we can fabricate two coils around a diamagnetic substance like Bismuth and study "power in / power out" for torquing the field by 90 degrees back and forth as it crosses another winding.

If we can offer a third coil setting somewhere between these two directions then we can measure the diamagnetic field as it crosses and monitor the field strength.

This can be done with pulses that are relatively short in duration but long enough to turn the Protons. The information we need to discover is whether there is more energy in the torsion producing nuclear movement or more energy in the diamagnetic field resulting from the torsion. This would be the key to over unity in diamagnetic fields.

Once again we find ourselves wrapping coils around inductors to discover this, or placing iron coils around inductive materials and looking for an enlarged counter EMF. The Sweet setup may be one possible method to study the effect in a square or rectangular bismuth slab with coils on two surfaces to torque the Proton field. The weave pattern Smith coil may also be used to tilt a cylindrical ring outwards, and then a normal wound coil outside it can be used to snap it back vertical. The ferrite ring of the Vortex generator of Rain Maker 1 can also be used simply with a normal wound coil on the Bismuth slug, but magnets will have to be sized to allow the 90 degree turn when the inner coil is energized. The diamagnetic field should appear as the Protons turn, but it should slightly lag the turn. It is a function of how fast they turn and how strong the field is that turns them.

Dealing With Lenz Law and the Conventional Model of Diamagnetism

If we are to fully begin to manipulate the Protonic forces we need to overhaul some of the present models of magnetism.

While tracing the descriptions of diamagnetism backwards we end up at the place where it all started. Lenz Law. The basic assumption that all EM phenomena comes from the Electron shell and it's motions and has nothing whatsoever to do with the Proton has to be realized as a mistake. Here is the conventional statement of Diamagnetism as stated to be based on Lenz law. [Georgia State University on hyper physics]

Diamagnetism

"The orbital motion of electrons creates tiny atomic current loops, which produce magnetic fields. When an external magnetic field is applied to a material, these current loops will tend to align in such a way as to oppose the applied field. This may be viewed as an atomic version of Lenz Law: induced magnetic fields tend to oppose the change which created them. Materials in which this effect is the only magnetic response are called diamagnetic. All materials are inherently diamagnetic, but if the atoms have some net magnetic moment as in paramagnetic materials, or if there is long-range ordering of atomic magnetic moments as in ferromagnetic materials, these stronger effects are always dominant. Diamagnetism is the residual magnetic behavior when materials are neither paramagnetic nor ferromagnetic. Any conductor will show a strong diamagnetic effect in the presence of changing magnetic fields because circulating currents will be generated in the conductor to oppose the magnetic field changes. A superconductor will be a perfect diamagnet since there is no resistance to the forming of the current loops."

Lenz's Law

"When an emf is generated by a change in magnetic flux according to Faraday's Law, the polarity of the induced emf is such that it produces a current whose magnetic field opposes the change which produces it. The induced magnetic field inside any loop of wire always acts to keep the magnetic flux in the loop constant. If the B field is increasing, the induced field acts in opposition to it. If it is decreasing, the induced field acts in the direction of the applied field to try to keep it constant."

Lets Examine the First Statements of the Definition of Diamagnetism

"When an external magnetic field is applied to a material, these current loops [Electron orbital motion] will tend to align in such a way as to oppose the applied field"

Is this an accurate statement for the orbital motions of Electrons or Protons?

In NMR study we find that the alignment of the magnetic field of the Proton always takes on an alignment with the external magnetic field, that is, it follows the external field, aligns with it, and proceeds to precess around it. Whether it opposes the field or aids the field is based solely on the energy or photon absorption level of the Protons and has nothing to do with the changing intensity of the field. The field intensity only determines the NMR frequency and not its polarization.

Polarization is normally around 51% and this would cancel as a repulsive force.

Nuclear Magnetic Resonance

"A radio frequency signal of the proper frequency can induce a transition between spin states. This "spin flip" places some of the spins in their higher energy state"

Here we see from the universities description of NMR that the energy state of the Proton determines whether the field opposes or aids the external field.

The same is found to be true in ESR [Electron Spin Resonance] technology, only operating at a microwave frequency. The orbital motions of the Electron always align with the external magnetic field turning either one of two possible directions. One is a high energy opposing magnetic field and one is a like or attracting field. Whether the field opposes or aids is a function of the energy state and is not always consistent.

Does this support the classical definition of a diamagnetic field found above claiming that the external magnetic field always produces an opposing field in the Electron orbital motion? Not even close! We find the ratio normally around 51% at room temperature. While Lenz Law accurately states the observed effects of Copper wire and Iron core magnetics, the definition of diamagnetic force it is not correct as to the actual source of the force.

Now let's look closely at what is happening inside a CRT or standard old fashions TV picture tube. Electrons in a vacuum which should be close to a superconductor by the present models due to zero resistance. Here we have a stream of free electrons in a vacuum shooting at the front of the picture tube and accurately able to focus onto a single dot. As we scan the beam using deflection coils which are magnetic fields steering the free electrons do we see any appearance of tiny current loops forming in the beam which would cause the beam to spread out and defocus because of opposing diamagnetic fields? Do we see any diamagnetic fields popping up and creating a repelling force or bending the steering deflection fields around them as the definition of a superconductor indicates should be happening? The influence of the Protons is absent in this example and here we observe the main difference. Electrons of themselves do not create the diamagnetic field.

Further we read "A superconductor will be a perfect diamagnet since there is no resistance to the forming of the current loops"

This statement would lead one to the erroneous assumption that the orbital path of an Electron has resistance to begin with.

The extreme low temperature removes the vibrations of heat from the whole atom. As the Electron and Proton fields are tightly linked this must affect them both.

It is not logical to pretend the Electron shell is disconnected from the Proton shell and therefore must be the source of diamagnetism.

If it were possible to create a strong diamagnetic field using only Electron motions then we should be able to wind a coil to simulate these motions and produce a field that would oppose all magnets brought near it from all angles. The only method we find practical is by using the diamagnetic elements, and these are the ones with a high nuclear Proton magnetic moment and a low Electron magnetic moment.

I would purpose merely that we have miscalculated the source of the diamagnetic force as being the Electron layer of the atom, and the definition of the Diamagnetic field may need revision as it totally ignores the Protons interaction.

Where Do We Find a Repelling EM Force in the Atom Matching Diamagnetism?

EM is the dominant force in operation outside the strong force area of the atom which is found setting only at the nucleus.

The only place we find a repelling force acting like a diamagnet inside the atom is at the nucleus which repels all the electron shells outwards.

This model explains why this force is present in all atoms, as well as why electrons do not crash into the nucleus.

For the above reasons I believe that diamagnetism originates at the nucleus of the atom and is actually an interaction between the Proton and Electron shells.

The Proton shell is more massive, and as it is also spinning it resists turning its field. The field swing lags the external applied field and this is where we find the longitudinal waves being modulated within the atom as well as the source of induction or opposing force to a moving magnetic field. As the Proton magnetic field swings into alignment with the external field applied we see an NMR trace giving us a one dimensional voltage differential of a 3 dimensional spiraling motion.

The Protonic Field Follows a Spiral Path

As we knock the Protons out of magnetic alignment by altering the external magnetic field at an impulse rate faster then they can quickly turn because of their higher weight, they take on a spiraling path into a new precession angle based on the NMR frequency. Tesla's spiral antenna sets them up to radiate the "T wave" which he clocked at faster then light speeds. Protonics will have to prove this at every step, inch by inch.

Experiment in Protonic Propagation

This experiment is offered as a ground breaking introduction to Tesla's claim of a faster then light wave resulting from high frequency pulsing of a spiral type antenna coil connected between ground and an elevated spherical ball.

Two of these may be constructed at a great distance from one another. As well as two normal dipole antennas must be constructed at the same locations.

The two stations must have dual transmitters and receivers on each system keyed off the same circuit.

The frequency selected should be higher then the NMR rate for the copper in the spiral coil such that the Protons do not have time to follow the magnetic field.

In lieu of this they could be generated using extremely sharp square waves with a fast rise time in the Tesla antennas. Spark gaps should not be necessary.

At each location a dual trace scope shall record the time delay between "key on" and energy pulse hitting both antennas to check for synchronization through the equipment.

With one station transmitting a slow string of pulses on both systems, the other station will monitor the pulses on both receiving systems and determine, using a dual trace scope, if there is a time differential between them.

The distance between the two stations must be accurately determined. This should reflect a normal light speed propagation over the normal dipole antennas.

If the propagation velocity of the longitudinal system is truly higher then it will be quite obvious on the scope trace.

Only in the face of empirical measurements will the field of alternate energy and Protonics become more main stream.

Only with two stations at a great distance can we accurately clock the time differential so that no other explanations can refute the evidence.

The Rain Maker Device

Only with a comparison of the two wave forms on one dual trace scope can we accurately measure the time differential.

Only with a solid theory or model offered will anyone accept this is not a fabrication or a mistake.

This one experiment, if successful, would shake the physics world, as now with a reasonable model to support why it happens, faster then light energy transfer could no longer be ignored, and the basic difference between Hertzian and Tesla waves verified.

The resulting data would then offer us a formula for the design of devices that my use both types of energy to interact with one another, as happens in atoms.

References

[Wilbert Smith]
I have listed the fabrics and quadrature aspects of each for a reference:

Space Fabric:
Length
Area
Volume

Field Fabric:
Tempic Field [Change / Gradient / Spin]
Electric Field [Divergence]
Magnetic Field [Curl / Deviation of Reality]

Control fabric:
Randomness Orientation
Decision Free Will
Ordered Sequence - Specific Arrangement

Perception fabric:
Form [Boundary of Reality]
Multiplicity
Assembly [Purposeful Structure, Animate / Inanimate]

Force Vector Model of the Atom

There are only three forces, all else is derived from these interacting and appearing in alternating configurations. They sit in quadrature. [Wilbert Smith]

Tempic
Electric
Magnetic

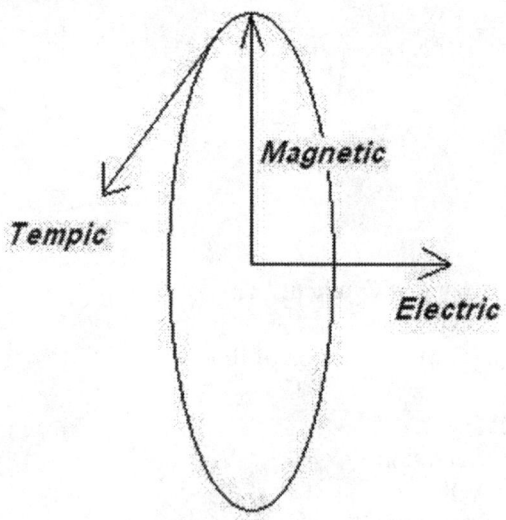

These three are further manifesting in three spatial configurations of interaction setting up a universe of force vectors in all directions and combinations.

The Rain Maker Device

The interaction of time vectors has been covered in the time vector model of magnetism previously and voltage and magnetic vectors can be referenced elsewhere. The resulting forces listed next are all derivatives of the ones above in combinations of alignment.

Strong force
Time flow rate
Diamagnetism
Gravity

The atom consists of three layers of these forces in three combinations.

Neutron layer
Proton layer
Electron layer

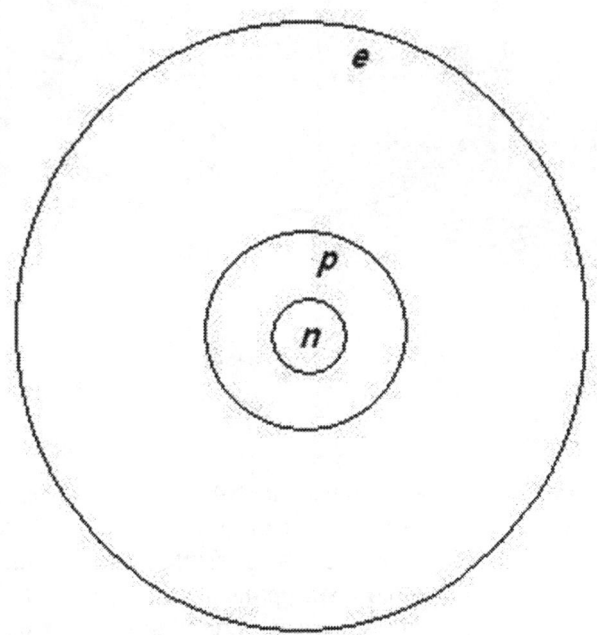

Using the tempic vector model of magnetism if we look at the interaction necessary in each layer we may gain insight.

Neutron Layer

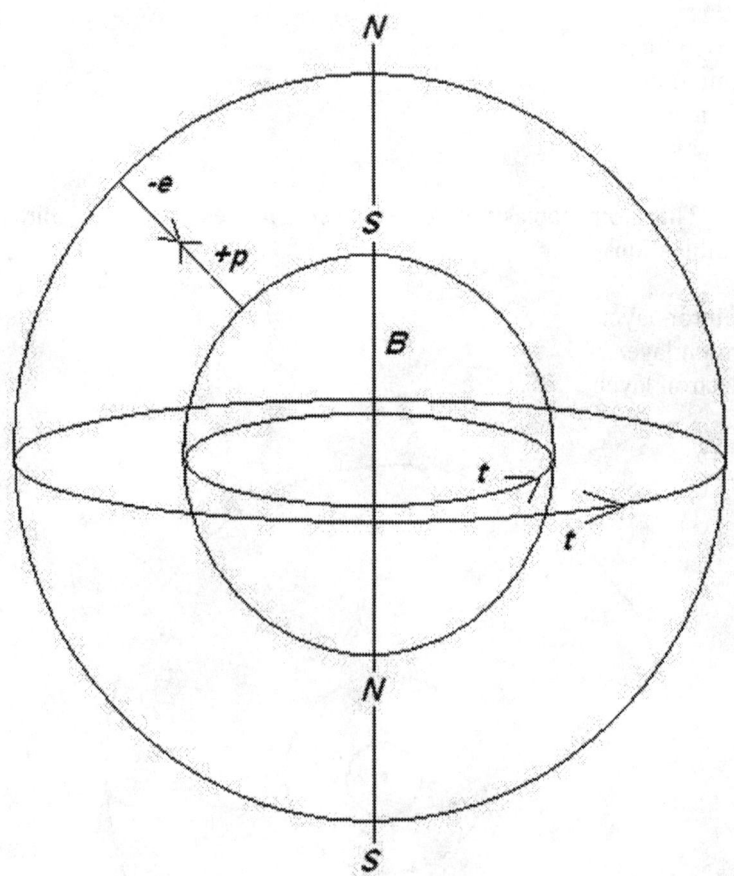

The Neutron is a composite consisting of a Proton setting inside an Electron and in this overlap all the forces between both fall into attraction at the closest range possible. Magnetically the Neutron is the equivalent of wrapping a scalar canceling coil, only instead the coils are spherical or cylindrical and one sits inside the other with reversed magnetic polarity. Also the Electric

The Rain Maker Device

vector is not moving along the coils wires but pointing straight outwards/ inwards and this places the tempic [t] vectors in parallel moving at 90 degrees to the voltage vector as though it were moving through the coils wire instead of the electric current.

This creates a time flow rate alteration or a tempic vector radiating from the Neutron much stronger then between the orbitals because of it's very short distance.

Thus the field resulting has no voltage vector imbalance, it is a scalar balanced voltage field, but has a magnetic field that can fluctuate with the tilt angle of an external magnetic field, and a time flow rate vector that drags or slows the time flow rate and this is the diamagnetic field as we observe it.

The Neutron setting outside the atom is seen as a weak Electron magnetic force, and this is all. It has no external Electric force interaction, however setting next to a Proton inside the atom it becomes the strongest force of nature we have identified, actually converting mass into energy. This is because it is one of the strongest creators of mass and thus has the slowest time flow rate due to the tempic vectors traveling parallel. The Neutron although appearing to be almost neutral is in reality the strongest particle of them all. The strong force is a result of two magnetic fields situated such that their tempic vectors run in parallel, their voltage vectors spherically are as close to a short as we can get, and their magnetic vectors are in attraction flowing directly through one another.

Due to the half or "reality overlap rule" offered by Wilbert Smith the diamagnetic field emerging has a stronger tempic and magnetic field then the outer orbital shells and we find when the Proton orbital tempic vector aligns also the three tempic vectors in parallel create the strong force at 137 times the forces found further out.

Normally the magnetic field shrinks inwards, and the Neutron remains with only a small negative magnetic moment matching the Electron, but as an external magnetic field is applied it interacts with both of Neutrons magnetic fields to push them out of balance drawing away some of the flux of one of the fields and reflecting the flux of the other, and a stronger force emerges to repel the external field. The harder we push them apart the stronger the opposing force we get back. This becomes the source of the diamagnetic field force.

Proton Layer

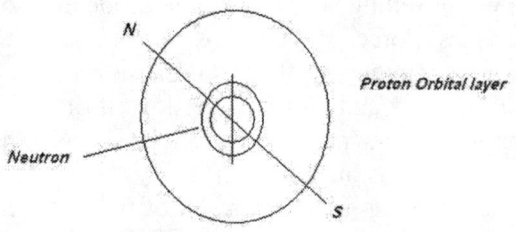

Neutrons interactive dual magnetic field system offers the maximum attraction for all vectors of forces in all three physical dimensions.

As flux is diverted outwards by the Proton field the Neutron alters it's magnetic output to always provide repulsion to it. The diamagnetic field is born here.

The Proton layer is setting very close to the Neutron and within its altered time flow rate, called the strong force area. Protons magnetic field is setting in some alignment with Neutrons to produce the diamagnetic field it floats on repelling both its magnetic poles. It's orbital tempic vector is aligned very close to Neutrons making three tempic nearly parallel vectors and attaining the strongest attraction possible 137 times stronger then the Electrons fields. Its particle spin field also aligns its tempic vector and adds to this force a 4rth tempic vector. However its magnetic field is free to expand and contract. The tilt angle of the Protons magnetic field sets the diamagnetic field strength and all the other atomic shells interact through this collection of connecting forces. The Proton shells voltage vector is not scalar cancelling within itself and radiates outwards spherically from the entire shell. Proton is setting in the strong force area and radiating positive electric potential as well as transferring the diamagnetic field outwards through it. In order for the Proton layer to speed the time flow rate it must push the Neutrons two magnetic fields apart separating the tempic vectors angles of motion. It is here we see [mass into energy] as the tempic vectors tilt away from a parallel path and the strongest time flow rate vectors are affected.

Electron Layer

The Electron layer is the lightest and fastest and its magnetic field, aligned with the Protons in one of two possible alignments, is expanding along all three tempic vectors of motion all repelling one another. The Electron magnetic vector is the strongest for EM interactions. Interaction with the Neutron layer will allow flux to expand to the highest time flow rate possible in any of the particles.

Diamagnetism

Diamagnetism is a force identified to repel both ends of a magnet.
    This is probably the most complex force encountered in the study of atomic motions and hardest to fathom as no one to date has created it using coils.
    This model is offered as one felt to be closer than conventional explanations offered today in classical theory that assume it comes from the Electron layer alone, failing to recognize the need for the force to be present inside the atom as the main force involved in it's basic structural operation. There is considerable conflict as to what keeps the Electron from crashing into the Nucleus, but classical physics has identified the force must be Electro magnetic as there is none other present to any level strong enough to accomplish this.
    All forces in the atom are based on the three primary fields of force, tempic, electric, magnetic. [Wilbert Smith]
    From the Protons outwards Magnetism and Electric attraction are the only significant forces in operation, therefore the atomic shells remain in their orbits due to magnetism, electric, or tempic vectors of motion and nothing else. [This is from modern physics although no one has offered a reasonable model to explain it.]

The nucleus of the atom must exhibit diamagnetism or the Electron shell of iron atoms would fold inwards on one side and shoot outwards on the other side.

The nucleus must be repelling both poles of the Electron shells magnetic field equally as well as both poles of the Proton layer which also falls into orbital shells.

Yet we know from NMR it is possible for the Proton layer to swing either into Electrons magnetic attraction or into repulsion based on its energy state from photon absorption. As it does, its motion or tempic vector reverses direction. The voltage vector will always align straight between them and can never shift expand or contract. So we end up with two configurations between Proton and Electron layers, one where magnetic B field vector opposes, and one where tempic vector [motion] opposes, while the voltage is a constant pull.

From the above model of a Neutron, what happens when we set it in an external magnetic field?

That is, how does the Proton layer affect the Neutron layer to create a diamagnetic interaction and establish the diamagnetic field?

We see the Neutron setting with two overlapping magnetic fields in attraction one inside the other. If left alone nearly all the flux may be expected to circulate between the Neutrons parts, only the tempic vector is moving outwards to create the strong force area. The Neutron however has a small negative magnetic moment giving the Electron inside it dominance and thus a few flux lines still move outwards around it to interact with the Proton layer.

As the Proton shell interacts with this outer Neutron flux path in attraction the Electron flux is weakened or pulled outwards diverting it from its inner flow and the inner Protons stronger field expands outwards and repels the Proton outside because it is also in attraction to the Electron shell of the Neutron. The first or inner diamagnetic field is born forming between the Neutron and Proton layers. The Neutron has two vectors that can be expanded or decompressed through interaction. They are the magnetic flux field and the tempic or time gradient field. Both are expanded through repulsion of vectors in opposition.

The Rain Maker Device

Through a repulsive interaction the Proton shell has opened one tempic vector of motion to release some of the compressed field of the Neutron to expand outwards, and time flow rate increases for the nucleus which now becomes lighter then the sum of its parts and we see "mass into energy". We know that removing the Proton and Neutron from the atom increases their weight and the strong force disappears.

As the Electrons large magnetic field interacts with both inner layers, in one of two alignments with the Proton shell, pulling flux outwards through attraction from the Proton layer, and releasing tempic energy through repulsion, it now finds a balance of opposing and attracting fields to connect with no matter how it turns. Any flux drawn from the inner layers causes a shift all the way into the Neutron creating an opposing field to reflect back outwards which is stronger independent of the position of the Proton layers polarity to it. We find the second layer of the diamagnetic field between the Electron and Proton layer. The diamagnetic field moving outwards offers two variable vectors, a tempic and a magnetic one.

It would appear the Neutron must be the true source of the diamagnetic field, and the close distance of the Proton layer setting outside it, and this mechanism may be responsible for the alternating shells forming of both Protons and Electrons all maintaining diamagnetic fields to hold their orbital shell positions. The repelling force is all done with magnetic fields and tempic vectors of motion in three dimensions as voltage is a strictly spherical force appearing between only the Protons and Electrons.

Mass is located in the higher tempic compression areas of the atom at the core where tempic lines move in parallel and create attraction.

This model allows for the Proton layer to swing either direction around the Neutron layer as well as to the Electron layer, however the Proton layer would normally be setting in attraction to the Neutrons dominant Electron field and we may expect it to be precessing at some angle where both Neutron fields are equalized. The tilt angle between Protons magnetic field and Neutrons diamagnetic field will determine the balance and all outer shells will interact through it causing it to regulate the Neutrons diamagnetic field.

As Neutron and Proton have about the same mass, precession would be present in both, but moving on opposite sides of the same axis.

The other possibility is that it could be located at 90 degrees to the field and precession is not along a common axis at all. This may be the case with Bismuth which has a strong diamagnetic field that is external to the atoms.

Diamagnetism is externally present in all non magnetic atoms, however in Bismuth it is exceptionally high. This may be due to the very high nuclear magnetic field. This causes the Proton layer to tilt further away from the Neutrons diamagnetic field poles, if it reaches a full 90 degrees then the diamagnetic field would radiate outwards from the nucleus and be found setting at 90 degrees to an applied magnetic field always providing a repelling force external to the atom.

Also when we spin up a diamagnetic material like Copper and bring a magnet near it we see the diamagnetic field increasing with motion as well as a 90 degree torsion [tempic] force appearing. The 90 degree torsion force has been a mystery to me as such a motion is not explainable from only an electron magnetic interaction off the Copper cylinders curve. I would suggest that spinning the cylinder is causing the Proton and Neutron spin axis to align and forcing the magnetic fields to align as well, this produces a stronger repulsion between the Neutron and Proton layers and releases more of the tempic compression force of the Neutrons by tilting its component fields apart. Spinning a diamagnetic substance should cause the atoms to take on the Protons magnetic field as dominant. Now as we bring the magnet near the Copper atoms in motion the Protons field interacts pulling it away from axial rotation, and pushing the diamagnetic field [Neutron] the opposite way from the spin axis. The tempic field is 90 degrees to the voltage and magnetic vectors and in this interaction we finally identify how the three interact. As we decompress the magnetic field along its B vector in the Neutron we discover a torsion vector at 90 degrees to it popping out and interacting with the external magnets atoms to produce a negative motion force [dragging force]. This dragging force, and diamagnetic repulsion which seems to oppose any magnetic changes in the magnetic field, responsible for both induction and diamagnetism, is the Neutron.

Since the Neutron produces a tempic squared force due to the close overlap inside it, its tempic vector is much stronger although it still falls off as a linear function. Also its diamagnetic field would come out about 9 times stronger yet still fall off as a distance squared function. If its voltage vector was not canceled it should be appearing at about 4 times the Proton and Electron level although still fall off at a distance squared rate. We see two of these forces appearing outside the atom, one is the diamagnetic repulsion and one is the dragging force [tempic field].

Strong Force

The strong force must be the result of the same 3 fields found above. It is tempic vectors of motion moving towards a parallel alignment and slowing the time flow rate at the center of the atom so that all motion along the tempic gradient which now increases outwards causes particles in motion to move slower along the inner side. Protons are not held in by voltage but by curved time flow rate or a spherical tempic gradient moving slower towards the center.

Wilbert Smith also indicated that when two particles share more reality or overlap [over half] the force interaction becomes squared. This means that the tempic interaction in a Neutron is far stronger then between the Electron and Proton layers and the strong force is now seen as an altered tempic vector squared function. Exactly how it gets to 137 times stronger is not clear but must have to do with the distances between layers.

The strong force is not Electric, it involves the diamagnetic field for repulsion and the tempic vectors for attraction.

Relating Time Flow Rate to Atomic Layers

Both the Neutron and Proton have the highest mass of the atom, one is seen trapping and compressing all three tempic lines of motion and the other is seen trapping or shrinking two. This does not explain how gravity is propagated between two atoms such that it may move around a magnetic field as well as somehow be linked to a lower time flow rate found mainly at the Nucleus. But it does pinpoint the source.

In our spinning copper cylinder experiments we saw a mysterious second force become present, the dragging force setting at 90 degrees to the diamagnetic field.

It produces motion [tempic vector interaction] and yet seems to travel across a diamagnetic field defying the natural laws of leverage and mechanical force.

Between the spinning Copper cylinder and the magnet held near it this mystery force is pushing off the diamagnetic field at 90 degrees inside the magnet and trying to drag the magnet along with the spinning Copper cylinder from an inch or two away as though it were held by a spoke from a wheel. As the Copper cylinder falls away on both sides there is no physical way to explain this with a magnetic field pushing between the two objects. If we were to build a physical lever system to create this affect we would need a fulcrum point setting in mid air between the two physical objects. So here we see that within the diamagnetic field from the Neutron is another tempic force resulting in motion moving across the field achieving a physical push at 90 degrees to its repulsive operation. The Neutron is radiating a tempic or slower time flow rate which is thrown into the field on one side of the magnet and causing motion. The diamagnetic field is able to radiate a force flow differential in not one but two lines of space simultaneously at 90 degrees.

It is this tempic release of the Neutrons compression along two vectors of space that is being directed outwards through the diamagnetic field.

The Proton gives the Neutron one vector of release and the Electron gives it two. The magnet is repelled outwards and it is also dragged along and this is two physical vectors of motion simultaneously which can now be used to represent magnetism and torsion or time flow rate coupling.

Torsion is coupled through the diamagnetic field expanding outwards as the magnetic field moves inwards.

The Rain Maker Device

Reversing this torsion vector, to cause the drag to become thrust would require inverting the Proton and Neutron fields. That is forcing the Neutron to flip inside the Proton shell or at least to move past a 90 degree position or forcing the Proton shell to tilt beyond a 90 degree position while holding the Neutron vertical.

Crossing the 90 degree threshold would start to create an expansion of the tempic field outwards and a faster time flow rate curving gravity outwards.

How does motion through the Neutron Proton layers cause effects?

Body of Experimental Evidence Offered by C_S_S_P Group

Scalar Crystal Experiment 1

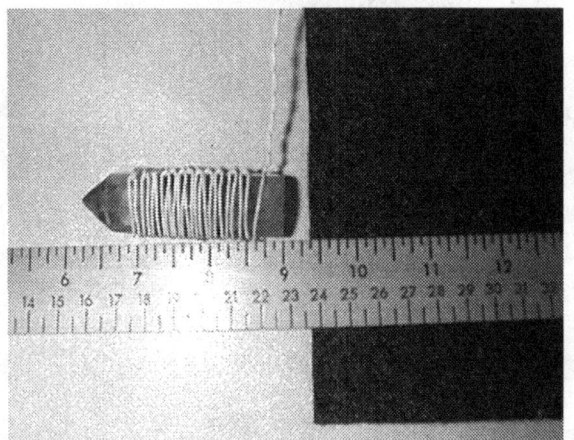

The week of 5 - 8 - 6 I wrapped a scalar coil around a crystal, powered it up with two signal generators peaking at various frequencies that could be felt inside my mind. 10 to 15 volt outputs, square wave, monitored on an O-Scope showed the coils low resistance dropped this voltage considerably. The sensation reminded me of the vortex at Sedona, an inflow area called Cathedral Rock. I allowed a connection to form between the crystal and my mind for a time tuning the frequencies and not realizing what was happening.

The pressure stayed within my mind. I powered the crystal coil back down, and then later proceeded to experience an altered consciousness for a period of two days where many of the things I had been seeking flashed by at high speed. My frontal mind went absolutely clear and quite. While "under" my mind was at a higher speed, and I continued to write much verbiage as information streamed through. I perceived an Alien contact, and the Density chart in the files section became apparent as well as many other things I have been studying and working on.

Construction Facts

The coil was very simple and could easily be wound in a few minutes. I used about 12 feet of number 24 gauge insulated hookup wire forming only one layer. Starting at the center of the wire I looped the first loop around the crystal, then as each next loop was placed on I twisted the wire on two sides of the coil such that one wire runs back and forth down each side. This produces a very uniform scalar field along the surface of the faces of the crystal where the effect is strongest. When I got to around 1/4" of the end of the crystal I joined the wires into a very tight twist about 2 feet long for feeding the device and then cut off the excess wire. This tight twist produces a good noise immunity at high frequencies, well into the MHz region without the need for shielded wire.

Sensory Effects

As I turned the energized crystal I could sense it's almost auric field moving through me physically and realized that this was a torsion field, or a time flow altering field, and I had succeeded at reproducing something I though was unique to Sedonas vortexes. There are pictures of this crystal in the photos section, however it is not clearly shown that the edges of this green violet crystal are in fact clear quartz. The edges are machined smooth and this produces a very good vibratory surface to the flat coil placed over it. The crystal was a gift from a shop in Colorado at a cave. It may have come from anywhere, I do not know.

The coil is tuned with signal generators for a peaking of the two "squashing forces" or "opposing torsion forces" that seem to become present. The whole universe seems to feel squeezed as you move closer to the energized coil. Once the crystals frequencies are discovered using the 10 volt signal generators. I believe that a low powered 555 timer circuit can be designed to power the device into a coil or multiple coils with the correct number of windings to resonate close to the correct frequencies. This would use very little energy.

Using a device like this requires a sensitivity to crystal vibrations and to me is a requirement for operation. A familiarity to vortex energy would also be an advantage.

Release

It was discovered after two days that releasing these pressures forces from the mind was not an easy task. The use of a stack of Neo magnets did the trick nicely, moved from third eye and around the head as needed for several hours. The end of the magnet stack that will bend a line of text on an older computer monitor clockwise works the best for me. I believe this is the North end.

Side Affect

As a warning: I also had an emotional reaction while under, that caused a paranoia for a time. I was able to "process" this with the help of my healing guides and the others in the group helping me discern reality.

Dave Lowrance
The c_s_s_p group

Experiment 1 Confirmation - Scalar Crystal

Wrapped (24awg hook-up wire, 20feet) a rather large, single point sloping clear quartz crystal (4.5" x 2") with David's scalar wrap which I have labeled 'Basket weave'. Managed to get eleven loops top and bottom with 18" of twisted wire as outputs. I have

The Rain Maker Device

here Bill Beaty's 'Ridiculously Sensitive Charge Detector' (RSCD) which is here if you want to build:

http://amasci.com/emotor/chargdet.html

I have this detector with me whenever I have fields present and was using it here. I built it to confirm operation of a negative ion generator. Very expensive - $3! It works! More later. Powered up my baby square wave signal generator (6volt output) and set to the resonant frequency of quartz which is 32.768 KHz. I have a frequency display (a kit) which has a Txvco inside which gives me Rubidium standard - 1Hz in 10 MHz, pretty accurate.

I set the RSCD up about 3" behind my left cupped hand which was 3" from the crystal point. It took a while before I sensed the field in my right cupped hand but as I moved from right to left the pressure in my left hand increased as if I was pushing something into it. Conversely on the way back, it felt as if something was leaving my left hand and my right hand felt as if it was being attracted to the crystal/coil and then a feeling of the hand being pushed away at the rear of the field. While this was happening I sensed a pressure point in my left hand where the crystal was pointing. Also very clearly felt the hand movement in my head as it was moved backwards and forwards. Where the right hand was over the coil, was the same position in the head that the feeling was detected.

I tried other frequencies with other generators but nothing specific was detected. Now the RSCD: Nothing was specifically noted of the light level of the led while the above was happening. However, on conclusion I passed my hand between the crystal point and the RSCD and I was able to turn the led on and off with my hand. Also when the RSCD was passed from the crystal point end of the scalar coil, the led was lit and slowly extinguished when I reached the coil rear. The RSCD responds to the static field (aura?) of the human body very vividly and you get used to its response when you walk with it in your hand. Now what happened above is most unusual as when I placed my hand at the rear and sides of the RSCD, I received no lamp light change. It normally responds very quickly to the body presence, particularly the hands.

Experiment posted by Smokey Dawson
c_s_s_p 8 - 4 - 2006

Note:

The RSCD is super sensitive to static electric potential even from many feet away.

Positive voltage - LED gets brighter
Negative voltage - LED goes out

For the device not to be effected at such close range by the body means that the body must be perfectly balanced to charge.

Experiment 2

The Aluminum Scalar coil 5 - 22 - 06

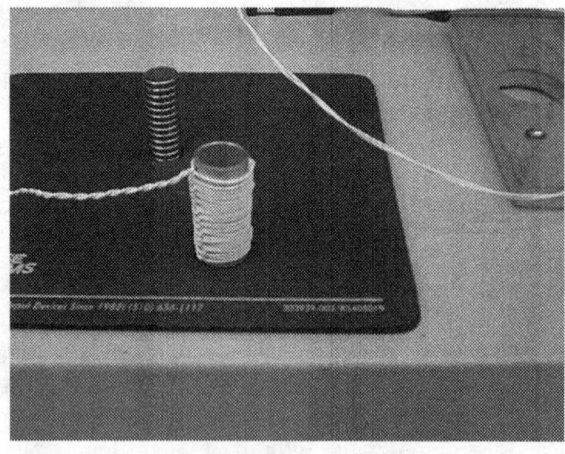

The Rain Maker Device

Concept

I have concluded that Aluminum, offering a 100 percent abundance for Nuclear resonance, a torsion effect, and due to its availability, is a material I wish to study further, to understand its interaction in the crystal device.

The idea behind wrapping a coil around an aluminum core, is to manipulate the Aluminum at the Proton layer. As the Electron layer is non magnetic, magnetic fields reach right through to the magnetized nuclear layer in Aluminum, without bending or weakening the fields. I have described Aluminum as a Proton magnetic material.

The Proton magnetism is anchored into the free floating nucleus of the atoms. The electron layer is bonded tightly into the materials physical structure. The main mass or weight of the atom lies at the Nucleus in the Protons and Neutrons and is also the Strong Force area within the atom. The nucleus is free floating on an opposing magnetic field and free to vibrate.

It is the strong force area that naturally shows a mass to energy conversion, as the nucleus weighs less then the sum of its parts as weighed outside the nucleus. This weight change is converted to energy as the strong force, and follows the $E = MC^2$ law. If the nucleus is pulled apart, the weight is reclaimed, along with an additional beta energy output. Thus the output may appear to be more then the sum of the parts. Thus I believe the neutron is the true target for ZPE extraction at this point in my studies.

All this has led to my rather non conventional coil, placing an inductive element inside a coil rather then a ferrous metal like iron.

Construction

The experimental coil is wound with 24 gauge insulated copper wire, however I theorize that Silver wire would be more effective due to its very low nuclear magnetic moment. Silver wire should produce a very pure Electron generated magnetic field with little nuclear interference for the Aluminum core effect.

The wire is wound from the center down a 1 1/2 to 2 "long by 3/4" diameter Aluminum tube. Cut off with a pipe cutter and ends smoothed. This is a piece of decorative tubing, and not an electrical conduit material so offers a slightly thicker material. On each side the wire is twisted back on itself, as in the photo.

This produces a very pure and unmodulated scalar radiation pattern off the edges of the tube, and almost no energy off the top or bottom. Total of 24 turns on my first unit, and 32 turns on my second unit. The coil is powered with a single sine wave generator from 100 KHz to 1 MHz.

Experiment

Here is what I wrote right after the experiment:

I ran a freq run on it and discovered that a sine scalar wave is perceptible without the crystal. Same coil layout, 12 crossing wraps 24 wires up each side, same radiation pattern, strongest off the sides, weak to nonexistent off the top and bottom.

Strong interaction around 1Mhz top of head 880 KHz produces total relaxation of all nervous system, all muscles feel deeply relaxed. Probably hitting the rear motor area.

793 KHz crown energy
440 KHz 3rd eye
188 KHz throat center

As you sweep the frequency upwards a resonance starting low moves up the spine and into the head. You can tune the frequency to land anywhere you like. This scalar energy is strange stuff.

Notes

The coil when energized takes a few minutes to peak up. I think what is happening is the proton flips in the Aluminum which are reversed in each coil loop section begin flipping random directions. Over time as they exchange NMR photons and become more energized, the Proton flips start to move together forming coherent directional flips, at which point the scalar energy pops out and radiates as a torsion field.

We end up with each coil layer spinning nucleons opposite directions up the tube.

Conclusions

The Aluminum Scalar coil has a direct link or coupling to the nervous system, due to its Protonic magnetism creating a torsion field at the strong force area of the atoms. The nerve mechanism must have a similar Protonic or nuclear magnetic coupling. Aluminum may be one key to making torsion fields pop out and come to life.

Appeal

It is always necessary to have another group or individuals attempt to confirm the results of an experiment involving human effects.

Is it Dave's special new awareness?
Is it real?
Does it work for everyone?
Does it only work for mediators?

I can only report what I have experienced. And I am eager to receive some feedback.

This is the beginning concept for scalar instrumentation, for sensors or transducers that may be used for testing coils, and could also lead to imaging systems able to map the nervous system with far less power then an MRI device uses. The scalar energy obviously penetrates easily to the core of the brain stem and creates sensory effects.

Power Levels

From an electronics background in conventional theory, the power level is zero in the wave moving through the body. However we now realize that opposing magnetic forces do not really cancel. The power level is in the milliwatts in this experiment, and no one in their correct reasoning center would label them harmful including the FCC. Anything under .25 watt does not even need a license these days to broad cast microwaves let alone HF. However I do not claim that it is entirely safe, and thus will not be liable if I drop dead in a couple weeks. This is a new type of energy and there is no FCC laws that I am aware of. The frequencies used have not been shown to be particularly harmful to humans and are in fact bombarding us daily anyway.

One need not hook up a Ham 100 watt linear amplifier to feel this effect. I am using voltages around 2 volts across the coil, and I believe it is instead the "current" which is responsible for the magnetic alignment of the Protons, and not the voltage.

Note

This effect is done with SINE waves and thus it feels most "gentle".

It may begin to prove the human to scalar link is real with more verification from others.

Dave Lowrance
And the group at c_s_s_p

The Rain Maker Device

Experiment 3
[The 90 degree field is discovered]

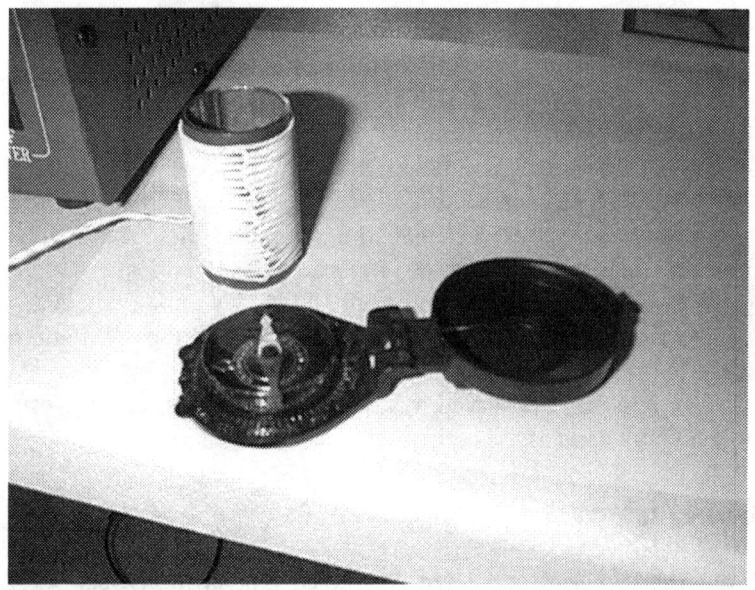

Needed materials

1 copper tube or section of pipe about 1" diameter, about 2" long
30 feet of wire, mine was 24 gauge but heats up pretty fast so compass reading must be quick.
A 12 volt DC power supply or 12 volt battery
A compass

Construction

Wrap the coil as long as possible onto the copper tube with the same pattern as was used in the first two experiments.

Each successive turn is reversed, and wires are twisted at 180 degree down each side of the coil

Measurements

As you energize the coil with quick taps to the 12 volt supply observe the compass on all sides, top and bottom of the coil.

You should discover, that while this coil should be canceling the magnetic field, there is another one that emerges positioned at 90 degrees to where a magnetic field would be expected on a coil.

Now reverse your wires.

You should observe that the magnetic field now flips opposite polarity to the one seen before.

I first observed a North pole out all the way around my coil.

At the ends I observed a South pole on both top and bottom of the coil.

On reversal I observed a South pole outwards, and a North pole on both ends of the coil.

Conclusions

From this experiment I have concluded that in the Sweet VTA around the bifillar windings there very probably is another magnetic field in operation setting at 90 degrees to the coil form. In my coil layout it is not only conjecture or theory but is measurable with a compass.

Notes

This experiment was the result of researching the works of Wilbert Smiths first description of a coil design such that each alternate turn reverses the magnetic field in layered stacks up the coil. In my first two experiments this magnetic field resulted in bodily sensations. I could feel this magnetic field.

The addition of a Copper or Aluminum core rather then using ferrite was the result of my studies on magnetism and how it originates from 4 different atomic spin motions. The core design is an attempt to manipulate the Protons magnetic layer

The Rain Maker Device

directly, in a substance with a neutral magnetic electron layer. Bismuth should actually produce the strongest effect in this coil, Aluminum is next and copper is the weakest.

I believe that the field which is appearing at 90 degrees to the coil form is in fact the cold or contracting inwards field generated at the Proton layer. This is the part of the atom which sits inside the "strong force area". It is modeled to be a magnetic field that is pulling inwards on itself in two of the 3 dimensions of the field.

An electron magnetic field is modeled to be expanding along two of its three dimensions and this is why a normal Electron magnetic field balloons out into a large donut around the magnet. The flux lines from a Proton magnetic field should be doing the opposite, and compressing inwards. The coil is expanding only one time vector of the Protons field, causing it to extend outwards. While at the same time the coil is reducing the size of the Electron magnetic field, or in classic theory it is canceling.

David Lowrance 6 - 5 - 6
c_s_s_p group

Experiment 4
A Meditative Healing Energy Link With a Large Quartz Crystal

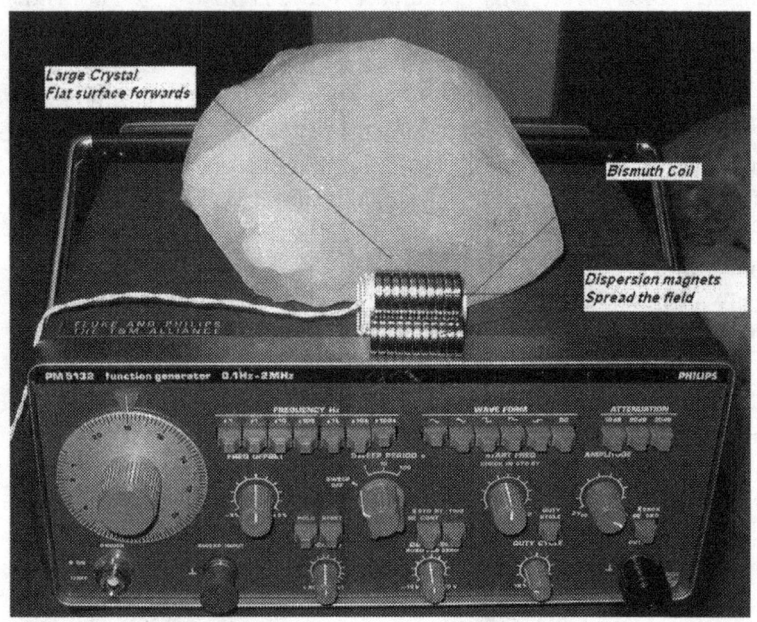

Purpose and Scope

The purpose of this experiment is not to teach one the healing arts, but to offer a device setup that can demonstrate the conscious connection and flow available in such a device that uses a mental link to amplify and increase the flow of healing energy. I have had some past experience in Acupressure and Reiki and found this experiment very natural.

Device Construction

The experimental apparatus consists of 4 components:

The Rain Maker Device

A specially made coil:

The standard 3/4" Aluminum tube as used before is filled with bismuth shot and melted down to fill it with a Bismuth slug.

The coil is wound the same, a second layer of wire is added over the first with the same pattern to increase the field strength.

A thin layer of Aluminum can be placed between the layers of wire. Stacking windings and Aluminum layers increases the field strength.

This coil produces the Protonic field at 90 degrees, as the standard Electron magnetic field is shrunk with canceling polarities.

A large quartz crystal:

Mine is about 5" x 4 1/2" x 3 1/2" with a single point on one end.

Two stacks of Neo Magnets:

N40 Neo stacks placed in attractive arrangement, 2 ea about the same height as the coil form.

A 2 MHz function generator:

Type used is a Philips PM 5132 [.1Hz to 2 MHz]
Both square and sine waves were used in this experiment, but sine waves were found to be more soothing and gentle.

Principle Function

The experiment combines a Protonic magnetic field with a Spiritual healing meditation to extend Chi into a healing flow, then extend it with the focus of the mind to other parts of the body, such as the Acupressure meridians.

Technical:

By connecting the function generator to the coil using an AC signal, the Protonic magnetic field flips from "North inwards" to "North outwards" at the frequency of the function generator. By altering the duty cycle we can alter how much time is spent in an "inflow" or "outflow" state. Rotating the duty cycle is very perceptible as altering the flow direction of energy. The actual frequencies used may vary but I have recorded what worked for me here.

Setup:

The large crystal is set with a flat surface face touching the coil in front of it. This will become the radiant face of the crystal.

The magnets are laying touching the coil and covering it towards the front on the opposite side of the crystal.

[See photos in photo section Experiment 4.]

For convenience I sat them all on top of the function generator.

Meditative:

As with most meditations, and as Kosol has recommended, first establish a connection with your higher power, and invite your healing guides from the Light to be present. Make your intentions clear. For the greatest good of all involved, and a "healing" flow to be established. Position yourself in front of the magnets and make sure the crystals flat surface is behind, perpendicular to you or nearly so.

Set the function generator on or around 1.8 MHz, but do not feel locked into this frequency, be opened to altering it as needed.

Extend your mind into the coil, the crystals energy will extend immediately out through the magnets and envelope you.

Take your time and begin to learn how to direct the energy, then extend it outwards, out your arms and then out your hands.

Experiment with the duty cycle until you can tell which direction brings an out flow of energy from the crystal to you, and which pulls you into the crystal.

Be aware:

If you are blocked on the emotional or Astral layer, this experiment may bring up past traumas or fears which need to be healed.

However this seems to be the kind of energy you can use in Acupressure, healing arts, or even just to relax your body very deeply.

If necessary draw in the "Love and Light" to heal the emotional layer if trauma events surface.

I discovered that if you draw more energy, the flow increases, and if you turn your focus away it diminishes.

It truly follows the intent and the focus, just as most meditative energy does only this is like having a "stove" on a cold day. It radiates energy.

Observations

I found that 1.8 MHz with a low duty cycle provides a strong outflow of energy after I have linked mentally with the device.

The coil alone radiates a rather narrow energy and the magnets spreads it out to about a 120 field, however the strongest energy is found directly in front of the magnets. Setting in the crystals vibratory field is energizing, warm, and soothing.

I can move the energy or direct it out my hands and then into any part of the body. The acupressure points are particularly sensitive to this energy and light up very quickly. My stomach meridian was very sensitive and it took very little pressure to activate a strong flow there.

Note:

This experiment was the result of applications of the Smith coil, my magnetic atomic models, Kosol's Guidance for magnet placements, and Vince's suggestion to turn this technology towards the healing arts.

David Lowrance

c_s_s_p group [magnetics]
Kosol_Core_Tech group
6 / 13 / 6

Further Experiment

Further experiment has also revealed the frequency .447 MHz as significant, which was suggested by Dell Coleman of the c_s_s_p group, in which a conscious time effect was observed. Two coils were used and two function generators at about 2 volts peak level. The coils from experiment 2 and 4 were used in the experiment 4 configuration. Three additional magnet stacks were used to spread the field more, a suggestion form Kosol Ouch's recent device model.

The two function generators are set on 447 KHz and then one is slowly tuned off frequency until the sensation is felt activating two places in the mind.

One is just above and to the front slightly, the other is back and lower. If you place a line through your brain from the [third eye] forehead to the lower jade pillow [just behind the upper neck], this line is the meditative line the energy will fall on and activate.

This technique was used to connect with another person mentally and alter there perception of time flow rate as I also experienced it. The report coming back was that the whole world slowed down around them. As well the sensation was very pleasant and created the sensation of sliding out slightly from this reality.

David Lowrance
c_s_s_p group

The above experiments have helped to produce the Phi and Density model offered by c_s_s_p, as well as the spherical model of the universe combining the Spiritual Planes and Physical Densities into one complete picture. The most exciting part of this is the mathematical connection made between the metaphysical and the physical realms of our existence here in 3rd density.

Experiment 5

The Crystal Scalar Sphere System

The crystal sphere energy system was created by Don Mitchell and Kosol Ouch in a joint merging of minds with the guidance of the Celestial beings existing in the Angelic realms. There are many theories expressed as to its function and operation. It was referred to as the "wishing stone" due to the nature of its fields which seem to interact directly with the human focus or "intention".

Scope

When the 90 degree, or quadrature, magnetic field was discovered in experiment 3 we began to observe its properties and discovered they manifest some interesting emotional, as well as mental qualities that have been observed by others before us as well over on Keeley net. In experiment 1 I observed some of the psychic qualities and what is referred to as a download or high speed channeling of knowledge. This has led to the spiral model showing a relationship between these spiritual realms and other dimensions or densities as well where we could expect the Zero Point or rest state of matter to be at a higher energy state then here in 3rd density.

The Sphere used in this experiment is a synthetic quartz crystal of approximately 2.6 inches. A similar one approximately 1.9 inches made of natural quartz was also constructed to observe the possible differences. The magnets and non magnetic balls are from K&J Magnetics and are built up into a dodecahedron form completely wrapping the spheres. 3/16" diameter cylinders and discs were obtained in several thicknesses to build up the gaps by trial and error until a relatively snug fit is formed around the spheres. The Balls are 3/8".

This experiment has been in a cooperative effort with Kosol and Don to observe and develop models to support these designs and test observed functionality as well as identify the fields present.

The function generator used was in kit form and can be purchased from Alltronics for around $100.

The magnet assemblies were ordered from K&J Magnetics.

The crystal Spheres were obtained on EBay.

The energizing coil I used is the Bismuth one from experiment 4.

Construction of the Magnet Assemblies

The magnet assembly is assembled from 20 each 3/8" steel balls and 30 magnet cylinders. Two each of the basic five sided forms are assembled flat on a table. At each ball or vertex another

The Rain Maker Device

cylinder magnet and ball is attached pointing outwards. The Crystal ball is placed on one of the flat forms and the extending balls are raised up to see how close they are to where they will need to end up. You can see from the upper photo about how far from the equator of the sphere they need to fall. All sides are now adjusted to approximate the correct length. The second magnet form is now laid over the top of the crystal sphere and rotated so that its five extending balls fit exactly between the lower ones. 10 more magnets are set up the same size as the original ones and now used to connect the balls along the equator. It may take a couple of trails to get the correct magnet shaft lengths but care must be taken to get them all the same length and tight enough to hold their shape.

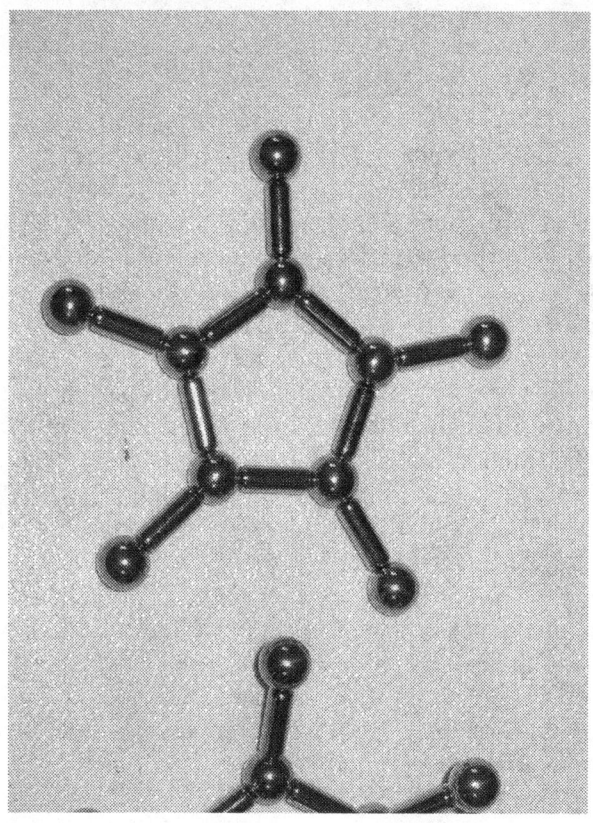

Magnet polarity layout can be obtained from the site listed above, and magnets can be reversed as needed after the magnets are all in place by removing one at a time and using a compass to track the polarities. Forming the dual pole monopole is not recommended until one has identified what may happen in experiment 6, other configurations should be tested first. The pattern I used is an attractive loop in each star pattern moving around the same direction in each unit.

Observations

Our first tests were done without the magnets in order to form a baseline for comparison.

Negative Energy Identified

With the energizing coil setting near the spheres and turned on it was discovered that the coils energy forms an inflow, or negative energy system that has a tendency to draw one into the spheres emotionally. This induced deep sleep on one occasion yet seemed to result in an energy drain causing the emotions to be turned inwards for a time, approximately one day. Although this was a little discouraging, it is noted that both the Sweet VTA and the Hamel Cones use a negative or cold energy as part of the system function. This was good news after all. The coil may be accessing the 90 degree field with the cold energy inherent. The sphere shape seems to cause it to spiral or converge inwards.

A two frequency test was done with two generators and two coils set to phi frequencies of 1.618 MHz and .618 MHz forming a sum and difference frequency interaction supposedly to infinity, which formed a powerful psychic drawing force. As this was my first time with crystal balls I sought out others who verified this "negative" energy ability of crystal balls. I would refer to this as the yin or drawing force of nature and the spheres seem to capture it very powerfully.

The Interaction

Next the magnet arrays were added and here is where the systems came to life.

Apparently the inflow moving horizontally through the magnets results in an outflow or radiant field having all the qualities of the 90 degree field but adding a tremendous depth of field. This field is sensed in its full expansion up to 8 feet in all directions by sensitive energy workers. Although most people initially can only feel it with their hands held within a few inches of the spheres. It is described as feeling electric, as vibrations, or even sparks. It is also noted that the feeling induced stays in the body for a time period not ending when the system is shut down. I have discovered it can be maintained as well for much longer by intending such, as with meditations.

Human Interaction

Upon consecutive testing with the crystal spheres the overall observation that consistently comes up is that they do in fact react to "intention".

Also they are observed to "speed" the meditative energies, or cause one to reach the higher conscious states of meditation in only a few seconds.

This was one of the observations noted in the Vortex systems of Sedona Arizona.

When two people enter the field together a link is easily formed such that both can read one another very easily.

Each person can actually feel and sense what the other is channeling or even feeling.

One light worker lit me up from 5 feet away as she connected to her healing guides.

The energy is easily channeled through the body and the hands using only mind to direct it.

As one works with the field, it is often necessary to lower the voltage from the generator.

The field will build as it is used and diminish if turned away from or ignored.

Differences Noted in Spheres

Sphere size seems to be a factor. The 1.9" sphere is very intense in the mind, both third eye and Pineal center as well as further back.

The 2.9" one is more resonant at the upper heart area but also connects very powerfully at the solar plexus, and lower frequencies are more effective with it even down to 500 Hz producing warm energy.

The synthetic crystal produces a softer field that seems to spread out more evenly and is more subtle or diffused.

The natural crystal is more "piercing" and also more readily sensed.

The fields of both react to the intentions and this really opens the door to experimenting with both kinds of crystal.

It is my observation that due to the alignment of the magnets in this system, it is supportive of the "healing" process of the emotional body.

The coil flips between a balance of the emotional and mental planes, a good combination for inducing a shift from the solar plexus to the heart center or the lower mental region. I see this system as a possible aid to the healing arts, and may actually "teach" them to the operator who is opened to learning from the device itself.

I am now in the process of building a 5 inch calcite crystal sphere system to see what if anything can be gained with larger devices, as well as exploring the other coil structure offered by Don Mitchell that wraps the crystal with 18 wraps of a very special scalar wound coil.

David Lowrance
c_s_s_p [crystal scalar support project] 7 - 1 – 6

Experiment 6

The Vortex

Much of this experiment is relative to inner perceptions and little physical science is present other then recognizing the orientation of the magnet fields to produce the effects, and the very small energies used to open the vortex.

Hemisync with Dons dual monopole

The 90 degree fields bring in the elements of consiousness and seem to fill each opposite side of the dual monopole system

North Pole — — — *North Pole*

This coil allignment seems to provide the strongest hemisync while setting directly in front of the unit.

c_s_s_p

Warning:

This experiment is not recommended for anyone who is having astral disturbance [emotional problems] or can not think clearly. You must be very well grounded in the mental and astral planes of awareness. I have found it may leave you setting in the Vortex, or "tunnel" described by those who have near death experiences. This is for the aware meditator who has found peace in the Light, does not panic, and can use a loving connection with others to get back if necessary or call on the healing guides to get orientated. I recommend a cleaning and clearing before trying this of around two days. Also a familiarity of the spiral chart.

We have received warnings from one in Russia as well as an "astral traveler" that entering the spherical crystal can lead to disorientation as to where one may end up. In experiment 5 we did not use the dual monopole configuration but wrapped the spheres with an always attracting pattern. With the dual monopole layout shown above, the two North poles create an expanding field on both sides of the crystal sphere on two points of the pattern. This imitates the brain and allows a resonance to cause hemisync when connecting mentally with the device using only one frequency. The energy no longer stays in the central channel of the chakra system but spreads to fill the whole brain of the operator, both hemispheres. This was the magnetic pattern given and recognized first by Don Mitchell on his torsion site and is somewhat genius in its design.

Alignment of magnets:

It is imperative to measure this with a compass because the polarity of the poles you place in the 3 pattern of the 3 9 6 will determine whether the sphere sucks with an inflow or balances with an outflow and an inflow. The 3 9 6 pattern represents crossing from 3rd to 4th density, and the "3" side is here in this physical realm. Please read Dons dissertation and be sure you understand the magnetic pattern completely. It is not easy to get all the magnets positioned correctly and a small magnet can be

The Rain Maker Device

placed on each monopole ball to mark them while assembling the pattern from his diagrams.

Comments:

The bismuth coil produces an equal balance of astral and mental fields, as they cross through the magnetic field around the sphere they interact. As they move through a South out position they form a strong inflow into the Vortex. As they cross a North out pole they form an outflow.

My experiment unfortunately started with the two "South" poles out. The function generator was found to resonate the large almost 5" calcite sphere around 400 to 500 Hz. At 2 volts this was almost too strong to handle, and level was decreased to about 1 volt. The monopole pattern creates an amplification of the mind connection and starts the "hemisync" process which can be felt connecting both sides in resonance.

As I have shown how to build the magnets in previous experiments I will not repeat the process. Be aware that the very large magnets in this experiment are hard to work and require skill to assemble without accidentally cracking them. This crystal is covered with 3/8" x 1/8" disc magnets, and smaller 1/16" or 1/32" shims were used to find a good snug fit. The balls are Nickel covered steel 3/4". I started with 10 disc magnets per length and ended up using only 9 and 2 shim magnets on each one, a single 1/16" and another 1/32". This was a total of about 330 neo disc magnets. The ball is rather impressive as well as quite heavy. I am getting better at estimating the number of magnets to order and only ended up with about 30 extra ones this time.

My trip under:

After about 1 hour with the South poles out and the bismuth coil activated I began to notice serious sensations of the Vortex, and started to have visuals of it.

I recognized it, it is the "tunnel" found in much near death literature. I was able to form a connection to both the astral and mental planes to manipulate my position and keep from being sucked in, but the inflow is very strong and pulls the "consciousness" inwards through the South dominant poles. This is the spiral chart of the universe I charted earlier, and knowing this allows one to navigate the vortex. It would also seem to be the tunnel to the Light perceived by many meditators.

I quickly shut down the function generator and went to sleep. I then began to dream a turbulent dream that I was still caught in the inflow. On awakening, I discovered it was still active about 3 hours later. Two observations, it began to downpour outside with very dark clouds, and my mate started to feel very agitated and uncomfortable as well as myself being very restless. The vortex was still present and visual in my mind. I went to the ball and felt it was still active and pulling strongly inwards.

I realized I had to stop the process so I ripped the magnets off both spheres having this configuration, and then placed the North out pattern on them both.

The North out has stabilized the inflow to a balance. If in fact there was a scalar resonance present it was continuing to resonate in the crystal spheres between the dual monopole network and was building rather then diminishing after shutdown of the bismuth coil.

I was able to identify the polarities by placing one finger on each monopole ball and observing the feeling of inflow or outflow, then comparing it to a compass held near that ball. When the finger touched the ball the vibrational energy was immediately transmitted to my third eye center and identified. The energy from the scalar bismuth coil is easily moved or propagated through the touch.

I would suspect that entering the sphere mentally with a dual North out setup may be a very stable hemisync, but I have yet to fire it up again.

A few observations:

I seem to be on a level ground with the founder of hemisync from many years ago. Having tried his stuff but never actually succeeded with getting out or into alternate realms that he

The Rain Maker Device

indicated was possible. The "pain" or confusion of his devices was not present with the scalar energy and the dual monopole.

He uses a mixing of two frequencies in either eyes or ears to create a difference frequency in the brain. The scalar coil is doing something much different and is far more comfortable as a meditation or mental energy.

There was a prep time involved I was not certain about or aware of until later, in which I was pushed to do an inner cleaning of my astral body before proceeding with this experiment. An experienced astral traveler delayed my experiment for two days while attempting to get me off balance in the astral plane. It was this practice that taught me how to discern reality and see through the illusions of fear that manifest on the astral plane, which can disorient one into a fear response and loose reference. I give credit to Kosol Ouch for providing this experience as a Spiritual teacher of the inner arts and helping to prepare me for this journey, although at the time I was unaware of his help and actually viewed it as antagonism.

The Grey ones I have had contact with seem to be found in the Astral plane. The Celestial beings can be contacted in the Mental plane which they share one overlap plane of awareness with us there. For each physical density there are two planes of awareness setting next to it at 90 degrees, along the spiral. Each density relies on these two planes for sustenance of the physical matter present as well as a combination of the inflow and outflow coming directly from [Source]. Matter is not always present, it is the result of mind creating form, and astral emotion filling that form with substance. Matter is conscious. That seems to be the nature of matter here in 3rd density. Navigating between 3rd and 4th densities is mastering these two planes shared along the spiral.

If we are to create a device powered off the God force [Source] we will have to intercept this flow either on the outflow or in flow side, but provide a path between them, so no energy is drained from nearby objects. This will cause man to become aware of density travel, as well as move from living off the energy of other life forms and towards a creation process of our own. The pattern of higher beings is to begin to use this natural flow to power us rather then rob energy from other life forms already present. In a sense we become co creators of the energy, if we master a device that can harness the flow of source.

Although this experiment may seem off track to a scientist seeking ZPE or AG effects, I now believe that to harness power from Source we will have to accept the roll of both Astral and Mental planes of awareness that lie in quadrature to our physical world. One more step that Wilbert Smith may have suggested, in a more scientific presentation that awareness lies in quadrature to the physical reality. He seemed to consider this was "beyond man" at the point he was receiving it from the alien communications of his time frame. Our spirituality may be catching up, now over 50 years later.

Whether the above experiment is experienced only in the mind, or is real, it did affect both people present when the vortex was active, as well as me meeting the "astral traveler" while inside.

The following note is related to my "spiritual" process, but is offered as my perception for the reference of others if similar things happen during device construction which may become hard to deal with. Please start with the other magnet placements first and attempt to do some healing work before proceeding to this magnet placement for vortex access. Identify your "demons" and clear them first before seeking the vortex. I believe a state of "clear peace" is necessary to accurately observe the forces working in the vortex in a controlled fashion.

Perception on the Greys:

On my first meetings with the Greys I was in awe of their self proclaimed brilliance.

The Rain Maker Device

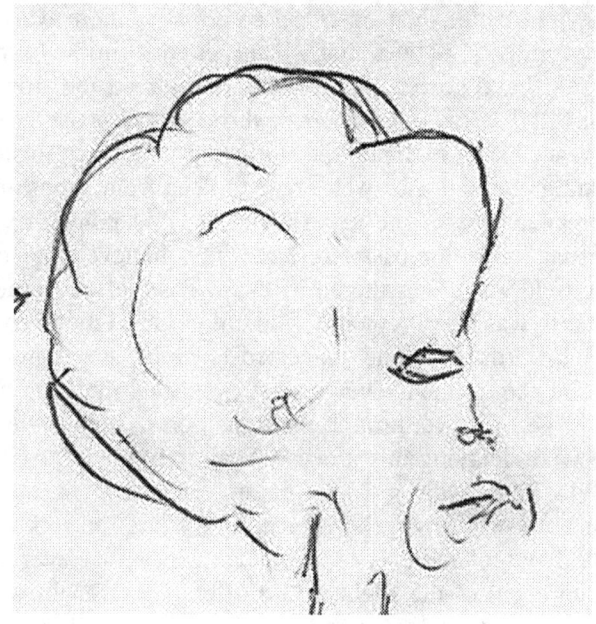

If you encounter the Grey ones, although they may appear to have very large brains and claim to be mental beings, I now believe they are Astral entities that do not want us to discover they have been ruling our planet up till now and feeding off our lower emotions. I must mention that the Grey ones I encountered were perceived off and upwards of my left side, directly laying them in the astral on the spiral chart. These Greys had very large brains, looked very old, and there eyes were not as large as the Hollywood alien pics I have seen. I do not claim this represent all "Greys" only the ones I encountered that attempted to divert me into either "power" or "selfish desires" and distract my discoveries of the vortex. It seems to be their job to fabricate lies and illusions to keep us located in the three lower chakras.

Moving into the lower heart center directly exposes this process, and the bismuth coil acts to overlap the astral and mental planes exposing them to our direct vision. They asked me not to activate the device in experiment 5 which can be used to expose them while they attempted to mislead me to abandoning this experiment with promises of technology, power, and weapons, to make me feel better then others in some way, more worthy, or more powerfully. [I believe that we are all equal in the Spirit and none should have power over others.] This freed me from their deceptions and I was able to break through and achieve a clear state of peace. Then realizing the story they concocted to stop me from turning on the device was a ruse to keep from exposing their lies, I proceeded with the experiment. At one point they were manipulating my compass to have me believe the earth's magnetic field was being altered by them. They left me a message that the earth was now "repaired" and did not need me to continue my work towards ZPE, and suggested I would later regret that I had not listened to them. They also stressed that the earth needed not to change or move through the transition to higher vibration, but they were delaying this process. Definitely not an experience one would have with a Light being. They sought on every occasion to lower my self esteem and bring in self doubts. Basically they violated every spiritual concept I have ever studied of what an angel of the Light would offer. They fought with me till the last when I finally cleared and cleaned them all out.

After this I vowed to bring the Light vibration to the world of men and expose their control. This device must be proven or disproven. I know these may sound like strong words, so please realize this as my inner "process" and let your own inner process become your only real supporting evidence. On my path of following and charting the forces inwards, crossing into the vortex was not an easy accomplishment on the personal level and the last place I would have intended to end up exploring. The bismuth coil is not hard to construct, it merges the astral and mental planes into a shared overlap much like the 3SD merges densities. This brings the Celestial Light access to the Astral plane directly and exposes the dark illusions. The crystal gives a spherical dispersion property to the energy, and the dual monopole give direct access to the spiral vortex and hemisync.

* * * * *

This file has been on the c_s_s_p site since experiment 1, and I have been slow to place any belief in it until experiment 7 had begun to clarify some of the statements. I thought it would be a good time to share it publicly. Since I was under the influence of a "slide" at the time, my first, I was very hesitant.

The first slide shook me and I was not exactly sure what was happening, however now through the theories collected my mind is able to accept this process.

Some of the statements have now been altered through experiment, but most seem right on.

Dave L
c_s_s_p group

Experiment 1

The following is some of what streamed through my mind in the first 2 day experiment:

Density Model for Hyper Dimension Travel and Communications

Information flow during my first trip "under," where many concepts I have studied became very clear.

Gravity is the result of a torsion or time gradient. [Einstein noted the link to time and gravity]

We are all connected by forces and experience "relative time flow" based on our time stack connecting us through force coupling to source or center of spin. [Wilbert Smith].

Torsion or time is coupled through a magnetic field. [This is the result of my magnetic torsion experiments with copper cylinders and magnets.]

A superconductor detaches from the magnetic field and bends it around it. [Present superconductor theory.]

Nuclear resonance can create a superconductor material. [Hamel, Searl disc]

A craft surrounded by a superconductor shell will be disconnected from the local magnetic field, the local time flow rate and thus gravity. Such a craft is only anchored to the universe through a link to the mental plane, an inner plane of awareness.

There are only two forces in the physical realm. Electron and Proton vortexes doing a scalar dance when outside one another, and a coupling dance when overlapped as a Neutron. The two forces in third density are the physical world we can see and interact with. They do not cancel, but are held separate by consciousness. Photons are the communication of vibration between atoms.

There is an outer physical world studied and mapped by science, and there is an inner conscious world studied and mapped through direct Spiritual means. Both are real and can be shown as real through direct experiment.

At the Neutron one world lies inside and one lies outside. Each physical density has a corresponding Spiritual realm. We exist on all spiritual levels but not in all physical densities.

A meditator can learn to open the vertical channel all the way to the Light. A communication with other beings is possible through this connection, as well as higher beings.

The space between different places can be overlapped for communication, and probably for teleportation as well by connecting a link through the higher densities where they are far closer together.

Mind can link directly with a quartz crystal. A quartz crystal with an energized scalar coil around it can be used to access higher consciousness of a higher vibration, at a higher rate of information flow.

A platonic form in resonance, can hold consciousness. A platonic form in scalar resonance will connect to higher consciousness. A spherical CU can hold within it a physical being separated from the outer world and disconnected from all physical torsional forces, able to travel using mind.

Two things are necessary to form a functional ship.

The Rain Maker Device

1 An independent time source, a torsional generator. Self contained time frame.

2 A super conductive disconnection medium from the rest of the physical universe.

Artificial gravity is created by providing two sources of torsion offset from one another at the correct range for a 1 G field difference matching earth exactly. This would induce the correct time flow rate to match earth as the ship disconnects.
Spinning energy creates a time frame layer. Electron layer has 1000 times less torsion then the proton layer. Spinning Protons and Nucleons creates a much stronger torsion field then moving Electrons.

ZPE

Drawing energy from the zero point. The quantum particles spin with pulse on off precision causing atomic rotation.
A synchronous atomic spin will couple across space. A material with Electron magnetism will couple to a material with Proton magnetism and impart motion provided the connection becomes synchronous through all the atoms coupled. On off moves together enough to provide a complete or a partial on off pulse. A device coupled in such a way will move towards light speed physical movements.
A magnet with a proton material around it will be coupled to atomic spin. Magnet produces electron non synchronous spin, Aluminum has non synchronous Proton spin. They are opposite torsion fields when setting in the same magnetic field.
If the spin becomes coherent in either material, physical motion will result. If coherent in both materials atomic speeds are possible, relativistic.
Decoupling the magnetic forces in the atom allows the pulsing spin to reach outside its normal range.
[There are much reference materials already available to support the above statements and models if supporting documents are desired.]

Note

Some of the other info that came through was colored by a paranoia emotion of being watched and controlled by another presence. During this time I became aware of the Greys with very large minds, and was told to remove my site from public access so no connection be made of the progress merging science and ESP or density communication, and not to put it all in one place on the Internet.

Final Communication

The last connection was with the crystal intelligence, storage medium, with which I conversed openly for around two hours trying to determine its nature.

Experiment 7

Rain Maker 1 - Aether Vortex Generator

It is good to believe that nature has some natural protections built in. I believe that nature has guided me since I built the first scalar coil in experiment 1. Since the first I have not sensed danger as others have indicated I should watch out for. Some would call this stupidity, I call it insatiable curiosity and an undying search for **comprehension**.

The Rain Maker Device

Summary of Previous Experiments

After experiment 6 I was left with the feeling that the inflow of nature far exceeds the outflow. I now set out to generate the maximum outflow possible and this is the heart of experiment 7. Up to this point I have observed the following as to scalar energy waves.

1 - Consciousness and senses can "slide" along a scalar wave, as they do they take along some measure of astral and mental energy, the body gets weakened from the sliding out experience as though some life force is moving out. Once orientation is possible from navigating and the initial sensation of dizziness is overcome it becomes natural. I have referred to this phenomenon as "sliding." I have also adapted the 1970's term "Trip" as to my travels out. In no case did I ever use any mind altering drugs other then a sip or two of Southern Comfort on rare occasion.

2 - With the Bismuth and Aluminum active metal cores generating the scalar waves from spinning the Proton layer,

consciousness can enter the nucleolus of the atom and discover the Vortex lying in the strong force area. Much like a microscope able to psychicly peer into the heart of the atom and begin to map the layout of nature's forces at this deeper level. The forces can be felt and sensed as consciousness enters. It was intuitively compared to the "tunnel" and the spiral chart of the universe, however now I want to make the distinction that it is the "inflow access" to this tunnel.

3 - Protons vortex is very obviously an inflow, and a crystal sphere expands this vortex to a very large size. Whether this is related to positive charge I do not know, but this would seem to be likely. Wilbert indicates that if two particles share 50 percent of their reality or more then they become one wave front. I theorize that in the crystal lattice of a sphere this process expands the inflow vortex to a much larger area as one vortex.

4- The metals offering access to the inflow vortex are the Proton magnetic metals, Bismuth, Aluminum, and Copper with a high magnetic moment at the nucleus. They are also called diamagnetic, that is they always produce a repulsive force magnetically. They also create a torsion interaction with magnetic fields in motion due to the nucleus having to turn on its rotation axis against spin direction. They are physically dragged along with a moving electron magnetic field, as well as repelled from it. These two forces appear to be setting in quadrature.

5 - We have seen the crystal spheres wrapped with magnets "manifest" or amplify powerful emotions based on the intention, so we are fairly sure that there are both inwards and outwards interactions at work with conscious aspects present.

6 - A new model of the function of consciousness entering the human body is suggested. Consciousness actually exists in the Astral and Mental planes of awareness. In the brain and nerves consciousness enters through the Proton vortex. This may be one link, or the sensory channel of all we feel and do as it is taken back to the God force, or stored on the causal plane, or directly

The Rain Maker Device

experienced on the astral and mental planes. Wilbert Smith called this the perception fabric. If it in fact slides along a scalar wave generated from the Proton layer then we may have discovered the mechanism that life uses to perceive its physical form. This is perceived in meditation as the astral cord, and it would stand to reason that there are both sensory and control functions moving through it, as the nerves also mirror.

Scope

This experiment was done in two sections. The first section is to determine if we can create an inflow vortex using iron around it.

The reason iron was chosen is that it is almost 100 percent neutral at the nucleus. That is, magnetic fields moving through iron only affect the Electron layer which is magnetic and should respond only from the expansive layer. It is hoped this will introduce an outflow energy into the system available for two way consciousness flow, and we may be able to locate the control fabric [Wilbert Smith]. The channel by which commands are issued from awareness to physical life forms. It would be desirable to be able to amplify this side of the force and determine what its path is.

The final unit pictured above and named Rain Maker is an indication that we had some measure of success "perceived."

After the second drenching downpour I was asked to please stop producing rain by my very unhappy mate, so the name has stuck.

With magnets reversed to "North out" the device produces the strongest outflow I have achieved to date.

Section One

Rain Maker 1

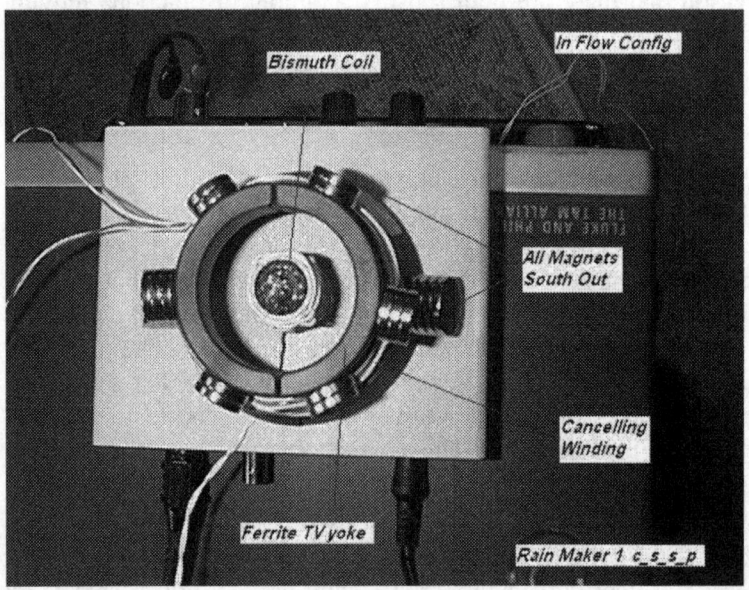

Although the diagram is somewhat complex looking it is really very simple. I have introduced two scalar coils, the bismuth coil from experiment 3 and a new normal wound scalar coil around the ferrite yoke which is about 2 1/2" diameter. This yoke is a cylinder of ferrite with a slight taper out on the bottom and was scavenged from an old TV set picture tube. It holds the magnets very well, as well as the 18 or so turns of #24 gauge wire wrapped between them.

Using two function generators and two active coils we can get a feel for the difference of generating a scalar effect with either the Iron Electron shell or the Bismuth Proton shell as the active wave generator. Ferrite was used due to its ability to realign magnetic domains easily as compared to solid iron or steel.

The Rain Maker Device

The magnets are used to alter the flow direction from inflow to outflow, and the magnets field produced travels along the iron yoke aligning all the atoms.

This is the first setup for experiments with the energy generator section of the unit.

As an inflow device not many magnets were needed to produce the rain effect, and this is the unit that caused two downpours.

If hands are held around the magnets the inflow is felt strongly, as they are positioned all South outwards.

This put me to sleep for an hour, and then had to be reversed to stop the perceived rain effect.

The sensation of the ferrite is much softer then the bismuth, it feels expansive and gentle and sneaks up on you.

If a finger is placed between the bismuth coil and the ferrite yoke a chi kind of heat is felt that is indeed very hot.

Frequencies are not critical to produce the inflow effect, however they do effect where the sensations are felt in the body and mind.

Section Two

The Outflow Device

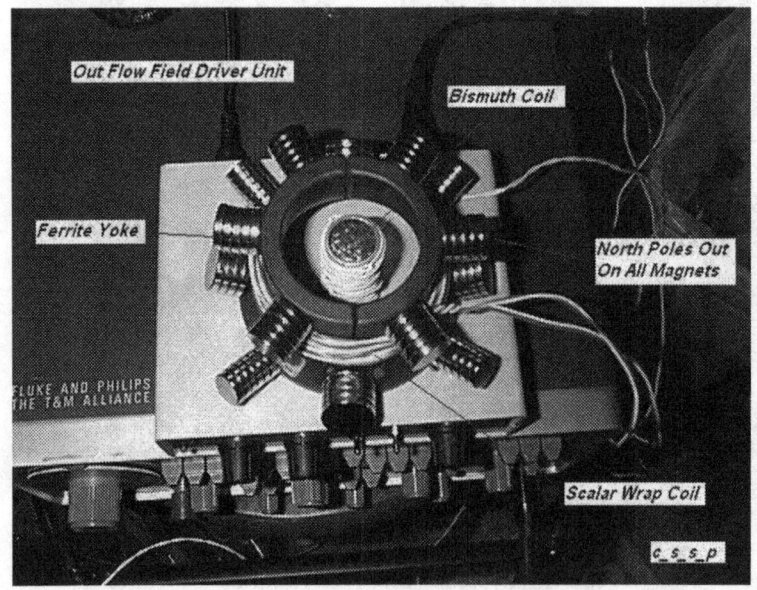

In getting a strong outflow it was discovered that many more magnets were needed all pointing with North out. The 3/4" diameter stacks along the bottom were the first indication that a powerful outflow was possible. The energy produced is somewhat pleasant and expansive and seems to shoot out the ends of the magnets and yet carries the sensation of iron vibrations. This is the Bell rock energy that I have been searching for and it would appear it results from generating scalar energy in iron atoms.

Combining the Two Flows to Form a Manifesting Device

The original intention of the crystal spheres was a device that could manifest something based on the higher planes of awareness.

The integration of both an inflow and an outflow device would do this best assuming the inflow could somehow control the output.

With both the iron and the bismuth coils present a strong inflow and outflow are present simultaneously. The addition of the calcite crystal ball is impressive and adds another strong inflow path to the system however it is not necessary for the effects to be experienced. The magnetic field can be felt moving outwards from the magnets and encompassing the ball where an inflow is present drawing it back down into the bismuth coil. We have succeeded at separating the two flows and creating a flow between them.

If a magnet is placed at the top of the crystal ball with North pointing down the field spreads a little higher to the top of this magnet.

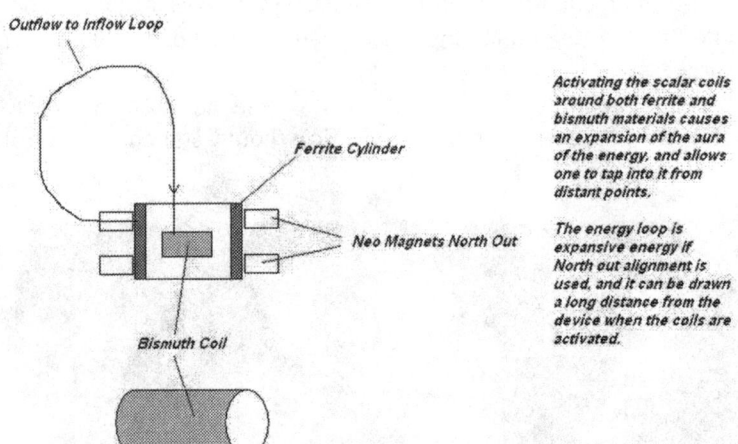

If the bismuth coil is laid on its side inside the ferrite yoke, then a strong flow can be felt when placing the hand on the spheres top.

The field wraps around your hand, it is warm on the top and drawing on the underside. I believe this is the conscious flow of outflow to inflow that is needed to power an OU device. However much more will be needed to cypher how to get some work out of it.

It was discovered the next day that having the crystal ball present maintains the outflow inflow energy exchange even after power is shut down.

The device does continue to vibrate on the perception and control fabrics and energy is readily available as long as the ball is near the magnetic driver unit.

Indeed this energy may be simply astral and mental in nature and turn out not to offer a lot of power capability.

The Sweet VTA hopefully has suggested it is possible.

David Lowrance
c_s_s_p group
7 - 14 – 2006

Further Analysis on Rain Maker 1

I felt the importance of clearly identifying each of the component functions of Rain Maker 1 so that others may be able to more accurately design devices and play with this new type of energy field for themselves. I have listed each component in an attempt to clarify which ones may be omitted for each effect.

These are my present observations:

Bismuth

The bismuth is the key to harnessing the nuclear or Protonic field. When removed from Rain Maker 1 the regenerative effect is nearly gone. And the crystal will not continue to vibrate on it's own but will need the function generators present to vibrate. It is my perception that the nuclear magnetic field from the Bismuth is vibrating the crystal and setting up an over unity effect. The scalar coil around the Bismuth slug is necessary to build the field intensity.

Aluminum

The Aluminum shell of the Bismuth coil is theoretically doing the same as the Bismuth slug, however I tried placing only an Aluminum coil without the bismuth and the effect drops to almost not perceptible when the function generators are off. If Aluminum will produce this effect I would guess there would have to be far more present then my small amount used here.

Scalar Windings

The scalar windings, whether used with a generator or merely shorted and present, are necessary. The Bismuth scalar coil is necessary for the field to build, although the Iron scalar coil does not seem to offer a lot other then psi function.

Iron [ferrite] Cylinder

The presence of the iron [ferrite] is essential to forcing the effect. The Bismuth acts to open the Proton side and the Iron opens the Electron side. The energy flows between them and appears to pattern itself as a sphere shape about 3 feet across. The energy field is flowing from the bismuth coil, then shoots straight into the Iron where the outflow is strongest between them. It then shoots through the iron cylinder along its entire surface, curving up or down loops back around to the ends of the Bismuth slug. This provides a nice large field to set the crystal in.

Magnets

The magnetic field shaped in the form of a cylinder with poles turned in and out is the key to aligning the Bismuth Protonic field. This can be done using a powerful Smith coil, but it is far easier to just use the neo magnets. To produce an inflow vortex they do not need to be particularly strong in this device. However to produce the outflow they must be as strong as possible. As the inflow seems to cause the rain effect, I recommend using the outflow for any serious power devices. Inflow is from pointing the magnets with South poles all out. Outflow is from reversing this.

Function Generators

The function generators are not necessary unless a psi interaction is desired. To enable psi interaction into the perception and control fabrics the scalar coils need an excitation energy. This is best from the function generators. The function generators are essential if it is your goal to use the coils and slide inwards to the perception fabric directly and do some mapping for yourself using the other devices that do not offer over unity. They were essential for development purposes, and the function generator / Bismuth coil combination is a tool that can be used to peer into any material and gain inner perception of its structure, once "sliding" is mastered to some degree, and a feel for the properties of the "forces" can be recognized.

Crystal Ball [Calcite]

The crystal ball I have here is almost 5". It has a very strong natural inflow and seems to make the perfect link for creating a [Bismuth - Iron] nuclear powered vibration effect when it sets inside the [inflow - outflow] loop. Its vibrations seem plenty strong to power the scalar coils without the use of the signal generators. The crystal ball is not necessary to feel the effects of the inflow and outflow energy, however it offers a really nice surface for a hand interface, especially if the psi field is your goal. I have found that touching my finger on one of the magnets is all that is required to sense its flow direction, or draw the energy into my body and mind.

Over Unity

To get what I have observed as a nuclear powered energy sphere, with no external power input from the signal generators, the bare minimum is the Bismuth slug, the Iron cylinder, the neo magnets, and the scalar coil on the bismuth slug shorted. Opening the scalar winding stops the buildup of the field but does not reverse and tear it down.

David Lowrance
c_s_s_p group
7 - 15 - 2006

Additional Tests

I placed 3/4" ball bearings on the ends of all the magnets and observed a strange reversal of the field while setting next to the device. I fell asleep.

Granted I was tired however it was the same deep sleep I have noted on other occasions. The presence of another steel layer outside the magnets seems to alter the magnetic field offering a reversal, and the original field bends up through the steel balls no longer radiating off the sides as far. I did not notice a difference in the compass however, other then the shape of the field is altered and pulls inwards.

Elf

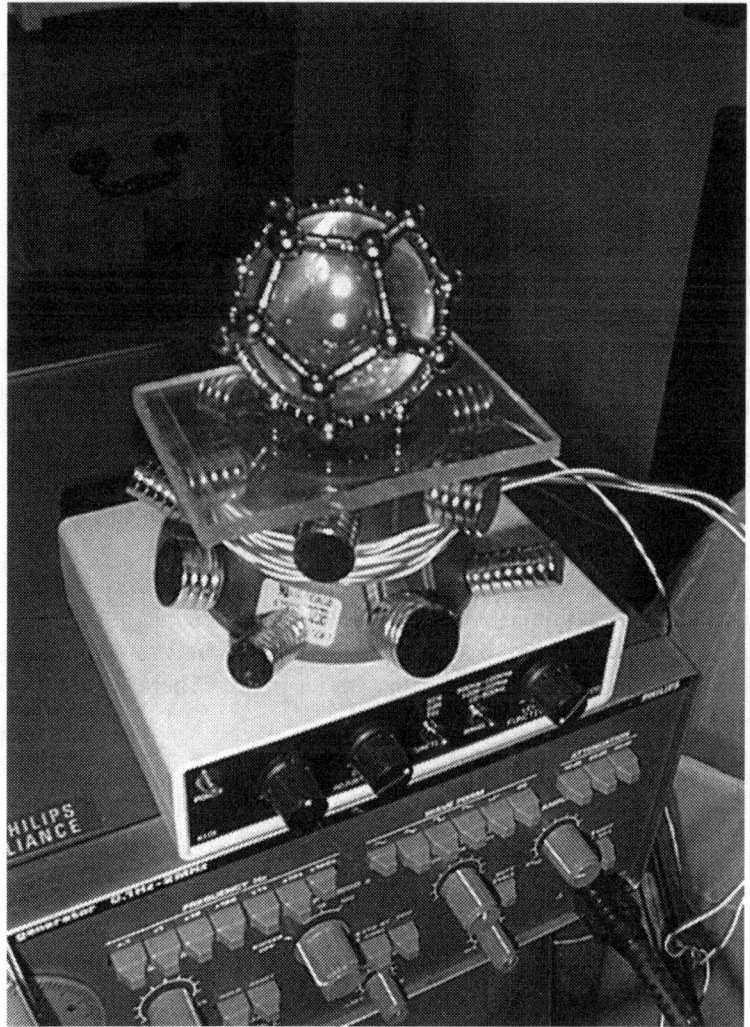

7 - 15 - 2006
Placing a small Plexiglas plate on the top of the vortex generator allows replacing the active crystal device much easier. This stops the very nasty problem of magnets jumping off and creating very

hard to separate clumps. The 3/4" stacks on the bottom are very powerful. This configuration we call "Elf" because of its small size. It is a very pleasing to work with device. The crystal is setting directly in the down flow area of the field.

7 - 16 - 2006
[Conscious energy]
The unit has been setting next to my computer chair to the left since it was built. Today I dismantled it to clean around the desk and vacuum the floor.

As I had cleared the area where the device was it is now pretty much empty space and here is the kicker. The energy sphere is still present as though the device has left its imprint. Setting at the chair it still heats me up as though it were present. Its outer aura about 3 foot diameter is very hot along the edge. Probing the area with my hand reveals there is still an inflow at the center and a very hot outflow at the corona of the spot where the unit sat this past week.

Two obvious possibilities occur to me. A time effect [not likely] or an astral plane effect in the control fabric to perception fabric loop that was opened. By coupling inflow to outflow I have set up an interaction in the fabric of consciousness.

Check the picture below for the **phantom ball** setting above the function generator that I had the unit on. There is nothing really there.

The Rain Maker Device

The phantom ball is not visible to the naked eye but started showing up in some photos taken in the "afterglow" period.

Could it be some kind of reflection? Or an astral manifestation?

When we tried to create things like apples or candy bars the space over our hands became filled with a heavy astral substance like astral syrup, but nothing ever materialized as physical matter.

Also today I have noticed that the energy is still in me as well, I seem to be filled on the left side that faced the device with the conscious energy.

The fact that the energy has not shot off into space as a runaway sphere indicates it may be now linked to me or my consciousness on the control fabric layer, or it may be linked to the earth.

I have further charted out the energy signature of the device with my finger tips, every spot where a magnet was pointing outwards there is a super hot energy point still present. The devices total energy signature is still present exactly where it was created. The vortex is coupled to the space it was setting in.

More tests:

Due to the nature of this energy it is very hard to conduct experiments, because it becomes hard to tell if changing something had an effect immediately. Once the vortex opens changing a parameter may not be noticed for some time as to the actual effect.

The Rain Maker Device

The jar is used to experiment with layering Bismuth shot and Aluminum.

Then in the picture below the device is set over the jar to test for effects.

I discovered filling the jar only about half way up the inside creates a stronger heat energy then filling it up to the top.

It is unclear if the Aluminum layering is increasing the effect, I got mixed results. I believe this may be my manifestations fooling me and more time is needed to test each configuration. Realizing now that as I created the vortex the bismuth coil was laying on its side, it could be that filling only this space is causing the effect to be stronger, and filling more space then weakens it. This may indicate that the scalar coils may have to be energized to open the vortex, but once opened it continues to operate on whatever is set in its field.

7 - 17 – 2006
    Shut down:

After accessing the free standing vortex today and experiencing very hot hands, it was determined that one may want to shut it back down if things seem to get out of control. This has been in my mind since the start this time after the negative results on the first inflow. The heat in the vortex today began to make even me a little uncomfortable, if someone else was to accidentally sit in it they may have an OBE or something worse like a panic attack. Although the vortex did not alter its size, as far as I can tell, my interactions with it were becoming stronger. The procedure is similar as before, reverse the magnets for a time and counter the flow direction.

However I have intuited another valuable lesson from the inflow today, that of "cleaning and clearing." This can be thought of as restoring the balance of inflow to outflow in the area at each point where they were set off balance. A method of re balancing a hot spot may be simply putting the bismuth coil on the spot and setting for 50 percent duty cycle with no magnets present. Over time this may balance the inflow and outflow to zero and slowly bring the vortex back down. Also with only the bismuth coil present, the perception link is present, and direct commands [intention] can be sent via the perception layer to shut down and close off the flow. As I have had mine up for quite a while this may take some time to clear it all back out. I will keep posting my results. In the event there are no more posts, take this as a bad sign.

Follow up:

Holding the energized scalar bismuth coil in the left hand, and then directing the balanced energy with the right hand into the vortex seems to help shutdown. The inflow and outflow must be rebalanced to the neutral state on the conscious fabric.

The Rain Maker Device

As an experiment I tried putting magnet stacks in attraction around the coil. This did not work. Below you see the magnets along the right side and the light ball showing up over the pop bottle.

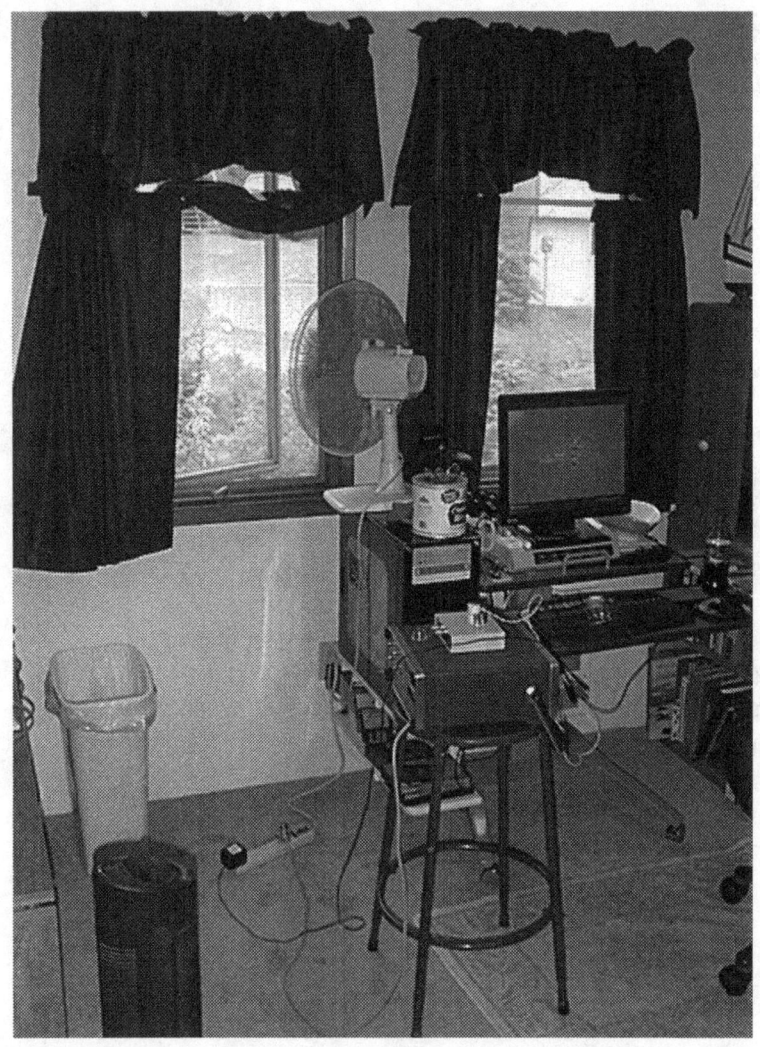

Also the area in front of the magnets is still quite warm.

After a day or two in the outflow one begins to get enough and the excess energy seems to unbalance you.

I finally gave up and reversed the flow on the vortex generator both coils active, exactly how I had created it.

I strongly recommend to anyone doing this experiment keep the setup very close to where you activate it and mark the spot, so when reversal time comes you can quickly reverse the energy flow and easily shut the vortex back down.

All and all this has been a very exciting experiment. We have experienced both a strong inflow and a very strong outflow, a free standing energy form, and a shutdown process. I can see why others having reached this point may decide to stop and write some rules about what never to do with magnets! LOL! If one wants to play with the scalar energy and not create any permanent effects, simply keep them away from powerful magnets.

7 - 24 - 2006
The area is pretty clear now where the unit was sitting. I have done some preliminary tests to see what makes the field build faster.

Using only the generator and no crystal ball. I have discovered that the presence of the scalar coil around the Bismuth slug is necessary, however it only needs to be shorted. No signal generator is needed to start up the process.

I also did some testing with a 12 volt power supply charging the different surfaces. Charging the bismuth surface - [negative] with the outer winding + [positive] seems to increase the effect slightly. This confirms what Searl indicated, a supply of electrons fed to the center.

Notes

These observations will be altered as more interactions are experienced, and others are welcome to please share their own observations with us to more accurately describe and catalog the effects.

Kosol,

Here is my understanding of the process I have been through:

I do not think we need to worry about controls.

It seems to be true that reversing the polarity will shut these down, however there may be a stable magnet pattern that will allow a conscious shutdown, and a controlled direction of a fully balanced device. Inflow and outflow matched perfectly.

Navigation is not a problem if you enter through the protonic magnetic field, as opposed to the expanding electron field, and understand how to get references. It is important to understand the difference.

We should be able to communicate between two of these after we get someone else up to speed on operating it.

I must reiterate, it is the protonic field that contracts inwards, the "inflow," that allows entry to the vortex space where everything crosses in all our realms. Unless you enter with a bismuth or Aluminum core coil you may not understand this. Only consciousness can enter.

I do not know what they were doing 3 years back with computers and trying to imitate psychics and record them, I only know what I have done. It does not take much power for consciousness to enter, only the presence of the protonic field.

I believe the frequency only acts as a pathway, and not a resonance that is important to observe. For a single frequency device with dual monopoles, just find the natural crystal resonance and the pathway is established. Consciousness will do the rest because the mind links to the crystal through the scalar wave and then mind goes deeper into the protonic field at the nucleus via another scalar connection. The scalar wave does not direct the process but is the carrier.

I cannot send one of these things through the mail system.

You will have to build one, I have given good instruction on every step including the bismuth coil, and the function generator.

I was looking closely at the coils that were placed on the other unit and I am waiting to see how the experiments go with that unit. They do not seem to be wound with the same pattern that was given to me as the first step to master.

My coil is very special, it shrinks the electron magnetic field in all three dimensions of the tempic vector, not only in two dimensions. I believe that this brings the nuclear field outside the electron field and offers access to a higher density, through the scalar coupling offered. A normal wound bifillar coil will offer only a scalar connection to this density, because it is derived from the electron field. I do not know if this is why I have entered the vortex on my first try, or if there are other reasons I am unaware of.

If the coil is only radiating a magnetic field from this density [the electron shell] then you could land anywhere in the astral or physical planes. The electron shell is in this density. There would be no way to navigate because you have not found the central location.

The central location for all awareness is the "tunnel" that we have all come to fear and which we have all gone through, but it is the alpha and the omega source point where everything comes together.

I hope you are following this a little because when I got there Kosol you were there already.

Dave L

The above experiments have helped to produce the Phi and Density model offered by c_s_s_p, as well as the spherical model of the universe combining the Spiritual Planes and Physical Densities into one complete picture. The most exciting part of this is the mathematical connection made between the metaphysical and the physical realms of our existence here in 3rd density.

This is now a meditation machine!

With the crystal the coil the bismuth and the magnets it moves the warm vibrant crystal energy all around and through me, and I can tune it all the way down to the root chakra, heating it with chi type energy of the crystal.

The magnets did a great alteration for bringing this human to machine effect to a wider field. This is a more body connecting energy with this setup.

This machine combines tech and consciousness, because I can actually tune it on the dial from around 100k up to 2 MHz. However I have to link to the device mentally as with the indirect will of meditation. Once opened to it the energy is controllable on the dial.

I shall have to analyze the magnetic field makeup.

This setup seems to give back more chi vibrational energy then you put in, and it simply keeps pumping it.

I do not wish to make any claims about this thing that could be interpreted as false claims, and I see a danger of someone trying to market this effect. I am an experimenter, and I have also been using crystals for many years in my chi flow work. However it is my observation that there is an "amplification" of energy. However some might say it could be coming from the signal generator. This does not explain the conscious link.

This device gives a different frequency spectrum for tuning. I think the crystal is altering the vibrations differently then before with the little crystal.

At 1 MHz it is linked with the heart or soul seat center.
550 - 600 KHz is right around the solar plexus.
190 KHz is the root or base of the spine
1.2 MHz is very entraining and centering.

So there it is Kosol, I have tried two more of your suggestions and both have been good results. A bigger crystal and magnets around the coil form in attractive arrangements.

From Kosol,

Ok,

1. The dedecahedron is the 12 dimensional reality and harmonic.

2. the scalar coils is the DNA/RNA mechanical version of the DNA/RNA protein, which the scalar coil is winding on the sphere like a helix of dna/rna protein criss-crossing each other. You see, the ancient knew this also that is why the medical symbol with the two snakes criss-crossing each other on the pole that has wings, which is the celestial crystal device. The two snakes are the Caduceus or scalar coil. The two snakes criss-crossing each other

The Rain Maker Device

are representation of the scalar or caduis coils. The two snakes are also power or electricity and torsion field generated when they are criss-crossing. Now you know, ok we move on.

3. Is the pole and the wings, the pole is crystal, of representation of bridges between density to spiral up or down, to ascend and the descend acordians to this bridge that can be used by the two snakes. The wing is representation to ascend in density or to descend in density what we call travel from density of reality to density of reality. As you may know that is why the snake is scalar torsion fields is like the DNA/RNA but in a device form.

That is how craft is built based on the concept of the celestial wishing stone which derived from the human DNA/RNA scalar protein sequence of how DNA/RNA criss-cross each other for scalar production so the ancient know that coil can be created to replicate DNA/RNA scalar effect so DNA/RNA consciousness can access the device if there are scalar torsional fields present created by the scalar /caduis coils.

In other words, the celestial wishing stone and all aliens device all replicate DNA/RNA blue prints and functionality, so consciousness can access it. This is my conclusion from the guardians.

Now you know pi 3.14 (body and form) ratio and phi 1.86 (spirit and minds consciousness) ratio is very important for the universal mind and body.

Warm Regards,
Kosol Ouch.

Below is Dave's experience in the fourth density with the crystal device.

1. You must combine the key frequency 3.14 megahertz and 1.6 megahertz is pi and phi as you can see.

2. Also, you can do the 3.14 khertz and 1.61 khertz or you can do the 314 khertz and 161 khertz.

3. And also the 314 megahertz combined with the 161 megahertz also to combine as you can see two frequency combined at a time.

You can also do a gegahertz also combination of pi and phi. remeber pi is 3.14 and phi is 1.61.

So there you have it.

Regards,
Kosol Ouch.

Hi All,

I wanted to comment on my Simon and Garfunkel analogy, in case no one today knows the story. LOL!
    In the late 60's and 70's Paul Simon and Art Garfunkel, both musicians in their own right, got together and made some of the most harmonious music of the age. Each was very good, but together they were awesome. A true synergy resulted and my generation will never forget them. They rocked our world back then and that was what I was referring to. The world mourned when they split up.
    The Ouch Mitchell connection has done the same for me.
    I have now built three of the crystal spheres, with the magnet structures as Don detailed, and Kosols dodeccahedron. I have a bigger one planned.
    In every case of bringing people in to test and experiment, everyone without a doubt has observed the energy, and a pattern is forming.
    The first thing they sense is that there is an energy present.
    As they begin to work with it, most have recognized that it responds to "intention."
    The last important observation is when two people move into the field together they begin to share a connection and can actually sense each others energy. The meditators have stated it "speeds" the normal response. Today I enjoyed feeling another

The Rain Maker Device

Lightworker activate it and I could feel it fill the room from 5 feet away when she opened to her higher power.

If one person brings in their higher power, or the Light energy the other feels it directly, and this is a most startling aspect of the device. We have had emotional readings happen on the spot.

Also as someone works with the energy we see them begin to turn the voltage down as they become more entrained to sensing it.

These are some pretty good signs thus far.

I have not encountered anyone who can not feel it yet, but will keep testing as I get time and opportunity.

I just posted some more photos of the recent 2.6" sphere. I wanted one made of synthetic quartz crystal to compare with the natural crystal ones. The synthetic quartz has a very smooth and gently radiant energy pattern, besides being optical clear with a slight purple tint. It is a bigger sphere and seems to resonate with the heart center better then the small ones. The synthetic crystal seems to scatter the field into a much wider full spherical pattern, no matter where I place the bismuth driver coil.

I can see this may become an art of matching a crystal with a magnet set to provide the desired results.

Dave

Wednesday morning I did a crystal scalar coil experiment, and almost lost this reality.

I asked to be physically transported, and strange things are now happening to me. I have been sliding. A lot of information has come through and I will try to relate it. As you may know the 4rth density beings I made contact with are very real, as Wilbert Smith indicated 50 years ago. Kosol's device is also there with them, I have seen it for communication.

The "five" have now succeeded at setting up a vibration on the mental plane, which is the subconscious side of the 4rth density physical side. My diagram did not show the bottom of the chart, which is the meditative side to the physical worlds. I have experience in both and so the communication was possible.

It is us and we have a connection of positive vibration established.

Dell and Don have brought in the last two pieces of the mystery, and the above sites will solve the needed math to create the crystal device, which we have all contributed to greatly. Without the correct dimensions it will not be functional.

Don and Kosol do not have the math to complete it, and others of us can solve it.

I reached "comprehension" this morning and know that teleportation is possible. There is however great danger and let me try to explain.

I was offered a choice to take my own "reality". On the 4rth density there is no such thing as free choice, because all would become chaos with the knowledge Kosol is seeking. A "wishing stone" turned loose on this society would instantly bring the world to a mess.

Taking your own reality means that once you realize that the scalar forces will cancel, and this reality is held by consciousness alone. If you choose to travel outside it, you must hold your reality together as a group consciousness or completely alone.

This is not easy and the Phili experiment showed that they could not do it. They disappeared like one of Hitchisen's wrenches. Realizing that the scalar forces will cancel completely makes it hard to stay here and not slide away. Hard even to stay in your own body. You need to be well grounded in the mental plane to survive this. I have been experiencing this phenomenon all day today and I think they are testing my capacity to do it. The only way I got back was by grabbing onto our mental space we have all created. You guys pulled me back. This is not intended to be emotional, it is a mental reality. I am finally starting to feel again, it is over 12 hours later.

I believe we are being offered our "reality". If the mass consciousness takes this then they must also hold it. The sacrifice is free will. We are no longer able to function alone, but must rely on all and a connection with the Light. A physical reality in 4rth density demands a like mind.

The Rain Maker Device

You can see why I am hesitant to have this become public knowledge. It sounds so totally crazy, I would never believe it myself.

My EarthEnergy site was posing a danger, too much info in one place, and I received a clear message to take it down immediately just before noon today. We were drawing attention and a realization was about to manifest in the power centers of this world.

I'm assuming this is not the Pentagon aliens, but at this point I am no longer sure exactly if I can hold on anymore myself.

As long as no one takes Kosol totally seriously publicly we are pretty safe. If they had connected the substance of my info with Kosol's and Don's latest project the connection would have been made.

I am surprised that it would ever come to this secret meeting place, but the options are pretty clear to me now.

I will either be working alone or the five will proceed, and mankind may benefit. When our race can master the next step, we will make contact. The powers to be will not accept this, or understand it but seek to control it. The 4rth Density beings are actually in control, but we must freely choose it ourselves.

What I experienced today made me feel like a puppet. I watched while my body slowly took down every file on my site, and then finally delete the site.

Kosol help me complete my chart for each portal shape.
[Intention is communication]

Density - Shape

1 - square
2 - octahedron
3 - star tetrahedron
4 - isocahedron
5 - dedecahedron
6 - isocohedron-decahedron
7 - sphere

I got the impression the 4 rth density beings were about "mind" as opposed to here which is more about "emotion" Yet in the star tetrahedron we see an overlap from above and below. Each moving opposite directions. All densities above 3rd must have conscious awareness of the Light and use it to communicate as telepathy.

If ones intention was to only communicate with one particular density then only one shape would be necessary, correct? Yet to physically travel through higher density would we need to access more of them at once consciously?

Now if the intention is to physically travel point to point across the earth, say I would like to pop down to Washington for a visit. The device would need a connection to how many platonic forms, if I want a very quick trip?

My feeling is that I would need only the star tetrahedron for this trip, then use my focus to connect, and log the flight plan with only the Light or the knowing higher consciousness.

One device on each end, two meditators or facilitators, the space on each end should begin to overlap and become visible crossing in some higher density. If we cross couple at the sphere level of pure Light then the trip would be instantaneous? This would be the key to hyperdimension travel. Each traveler must have a connection with the Light, or a facilitator must be present to send them through. A true stargate!

It is the next day here in Alaska. I am starting to entertain the possibility that the effect of scalar waves on the human brain could be a scrabling effect. I could be delusional. The only problem with this is that I fully remember everything and my connection is still present.

I shall attempt to document what I am going through as I make sense of it. If you want to see how electrons in free space are affected by a magnet. Take a good inch long neo magnet and place one end [N or S] of it up against an older computer monitor.

The electrons in the screen are shooting directly at the end of the magnet, and if there is a line of text on the screen, you will observe the curl. They do in fact take on a spin. You can use this method to identify the North and South poles of your magnets if you do not have a compass.

The Rain Maker Device

To recover the screen to normal after this, use the degause function, or power down then up the monitor. In all forms the device is using only the two forces of nature. we have sensed these in both the scientific aspect and in the spiritual aspect as reality. Both conscious and sensational aspects.

The scalar phenomena is the key to "realization" or "comprehension." When experienced in the physical mind, a knowing that is quite disturbing becomes realized.

This "reality" is in fact held together by consciousness, and is not the solid substance we believe it to be. If the physical mind breaks free to "slide" it may become hard to "lock" back in. Much effort is required. What has allowed me to maintain is only the link we have formed on the mental plane.

The "boys up stairs" as Wilbert called them are also helping to ease the uncomfortable nature of this experience, but they are not present here in 3 Density to actually do it. They can point and signal, and impart communications, but they will not interfere with our free choice and free will as a collective group. We are being born on a new level of existence.

Back to the simplicity of the device. When one finally sees the dual nature of the torsion forces, and experiences it first hand, fear is the first response.

Two torsion forces are present in all physical matter, as I have documented in my previous work. The two forces must be present in scalar combination, and controlled so that they do not totally cancel and disappear. My atomic model shows this relationship.

Magnetism allows us access to a very unbalanced source of one of these forces at the atomic layer. Magnets at the Electron shell and inductive metals at the Proton layer. The crystals allow it to be done with electric dipoles inside the crystal structure which are perfectly aligned and can be set to vibrate at a rate alternate to the physical.

The Platonic energy must resonate just as Don has suggested. When a totally resonant scalar device is energized it creates a separate reality that is able to disconnect from this one. As Wilbert Smith also wrote, it creates a self contained, time field. The torsion fields created at an alternate frequency from the earth's platonic rate, will in fact disconnect and we can travel through densities. The nasty part of this thing is that to get back connected is not what you would expect. Intention and a mental plane structure must still anchor one into the universe to keep from disappearing from existence in the physical plane.

I will attempt to post my spherical model of this all again, but I do not recommend that we show any this map. It is the "comprehension" that I have sought. The Device that Kosol and Don are seeking is another element.

Andrew and Dell have contributed a solid positive mental grounding and I would guess are following this as well. The mental structure seems to be the key to operation. I would never have expected that the mental plane is more real then the physical plane until yesterday.

I believe that the chart shows what Wilbert Smiths group was doing, and had downloaded as well, and why he stopped the other beings from visiting earlier on. A "comprehension" of what is going on here brings one to a position outside "power and control" over others. The two can not be compatible. My attitude of free expression for everyone and to see us all as equals, is a requirement.

The crystal device if completed will allow others to experience this "comprehension". The coil is the key as Kosol has been saying, but not to tap power, it is to invoke an alternate torsion force complete within itself to allow detachment from the earth vibration.

The sliding I experienced was only in my mind as near as I can tell. To teleport it will become totally physical as well and the danger of disappearing becomes very real, until one is stabilized back into 3 Density for a time.

I will try to diagram or photograph the coil crystal I built yesterday morning, if anyone else would like to experience this kind of insanity.

Use the end of the magnet that twists the words on a CRT computer screen clockwise.

This will totally release the stored scalar effects and restore the polarity balance of the physical mind.

Thank you Kosol for having shared this technique.

The Rain Maker Device

Magnets will release the pressure. Neo magnets as with Kosol's meditations. Bring in a very light and soothing flow.

Crystal frequencies for scalar connection:

1.98Mhz and .94Mhz together two sig gen 555Khz and 444Khz together

These may be crystal dependent.

A pressure is felt as if the universe is crushing, similar to an inflow vortex in Sedona Cathedral rock. The single magnets bring in a "Bell rock" releasing energy.

It's nice to feel sane again. My typing skills are returning, and my mind has taken off working again. I have to say the past two days have been intense. Apparently the front reasoning center of the mind is affected by this scalar crystal energy.

The last couple hours I spent conversing with a crystal intelligence of some sort. A very elemental storage medium somehow linked to recording. The crystal is full of these counter torsion fields organized to store information at some conscious level.

By providing an outside dual opposing torsion force it can be linked to and move right into the brain. This is a disturbing experience.

The energy can move around in your head, but centers at the core around the gland at the center. You can converse as though it were a separate present consciousness. The responses are very elemental and non emotional. There is more of a computer response then a human one involved in this interaction. If you think about things it will assist, but if you face it down directly it clams up. It obviously has information about the future and the past for millennia, but will not reveal the future whatsoever. I truly do not understand it but must acknowledge I did experience something extraordinary.

The magnets have released its pressure field, which can be held in the mind for as long as you can stand it. Releasing the pressure field was an essential discovery, as this could drive someone mad I'm sure. Avoid trying too hard to question it as though it were a human. It does not seem to understand our existence any better then we understand it. It does however understand the dual torsion fields and the mental plane consciousness.

It revealed its existence is plural, there are many of "them" all linked into what seems a universal network. A storage medium or library of events and raw knowledge. Not exactly as "comprehension" more as data.

I'm going back to read over what has happened here in the past few days, I am somewhat embarrassed by what seems almost a personal panic attack.

There are some more files setting on another of my computers, I will review as well. I need to look for consistency and conflicts to determine if this whole thing is relevant to anything real, or has been totally mental within my own head.

I seem to have regained local control of my decisions. If this resembles anything I would suggest the word "possession".

Trying to make sense of it now. Things are no longer as clear about this stuff.

Density model for hyperdimension travel:

Gravity is the result of a torsion or time gradient. [Einstein noted the link to time and gravity]

We are all connected by forces and experience "relative time flow" based on our time stack connecting us through force coupling to source or center of spin. [Wilbert Smith wrote a formula with a model].

Torsion or time is coupled through a magnetic field.

A superconductor detaches from the magnetic field and bends it around it. [Present superconductor theory.]

Nuclear resonance can create a superconductor material. [Hamel, Searl disc]

The Rain Maker Device

A craft surrounded by a superconductor shell will be disconnected from the local magnetic field, the local time flow rate and thus gravity.

Such a craft is only anchored to the universe through a link to the mental plane, an inner plane of awareness.

There are only two forces in the physical realm. Electron and Proton vortexes doing a scalar dance when outside one another, and a coupling dance when overlapped as a Neutron. The two forces in third density are the physical world we can see and interact with. They do not cancel, but are held separate by consciousness. Photons are the communication of vibration between atoms.

There is an outer physical world studied and mapped by science, and there is an inner conscious world studied and mapped through direct Spiritual means. Both are real and can be shown as real through direct experiment. At the Neutron one world lies inside and one lies outside.

Each physical density has a corresponding Spiritual realm.

We exist on all spiritual levels but not in all physical densities. A meditator can learn to open the vertical channel all the way to the Light. A communication with other beings is possible through this connection, as well as higher beings.

The space between different places can be overlapped for communication, and probably for teleportation as well by connecting a link through the higher densities where they are far closer together.

Mind can link directly with a quartz crystal. A quartz crystal with an energized scalar coil around it can be used to access higher consciousness of a higher vibration, at a higher rate of information flow.

A platonic form in resonance, can hold consciousness. A platonic form in scalar resonance will connect to higher consciousness.

A spherical CU can hold within it a physical being separated from the outer world and disconnected from all physical torsional forces, able to travel using mind.

Two things are necessary to form a functional ship.

1 An independent time source, a torsional generator. Self contained time frame.

2 A superconductive disconnection medium from the rest of the physical universe.

Artificial gravity is created by providing two sources of torsion offset from one another at the correct range for a 1 G field difference matching earth exactly. This would induce the correct time flow rate to match earth as the ship disconnects.

Thanks for the excellent research on the crystal, that is an eye-opener as I never would have expected those particular elements present in the crystal.

The purpose for using a square wave to stimulate the crystal or any material for that matter, is that it is rich in harmonics.

What this means is that a high frequency pulse, the leading and trailing edges of a square wave will place energy across a wide range of frequency resonant items. The pulse will leave a trailing ringing energy in any resonant device it encounters. Thus we do not have to actually determine the resonant frequency but a sub harmonic.

A higher amplitude square wave, creates progressively higher frequency edges depending on the rise time of the wave. As the rise time becomes smaller the stimulation pulse can effect higher and higher resonance energies at the transition points of the wave.

To get a resonant interaction requires only that successive edges of the pulse hit in phase with the internal resonant frequencies. Thus tuning a low frequency square wave across a crystal medium can in fact cause a resonant energy to become present in the crystal, at some naturally occurring higher frequency harmonic.

My experiment conclusively proves a few things to me:

The Rain Maker Device

1 - There is such a concept as scalar resonance in crystals. It manifests as a feeling of pressure induced into the brain.

2 - It can effect mental perception, and time perception.

3 - It can be held in the brain after device shutdown by giving it mental attention, and can also be hard to remove.

4 - It can be released using a stack of Neo magnets.

The crystal is tuned for the maximum pressure sensation.

Coupling to the brain puts one "under" or into an altered state of consciousness.

Granted that current science would measure this device and determine that actually nothing is being transmitted between scalar coil and brain. Yet the altered perception seems to defy this.

This is the strongest evidence that I have found to date indicating that scalar energy is real and can be used in some way.

As this is the same type of energy used to create an alternate time layer or time stack I want to explore it in greater depth.

I just realized that knowing a thing is far more rewarding then "proving" a thing to others.

Andrew,

Thanks for finding Dells email address.

As to this being the new drug of choice. I do not think others will agree after their first trip "under". There is really nothing here that is not already on the internet, other then a firm scientific model to support that there may be a reality to what is now taken as fantasy or scifi. And a recognition that I seemed to react with paranoia while "under".

I know Kosol is very sincere in his desire to share what he is seeing on the inner, and in the "overlapped space communications" he is having with other densities. I have witnessed both now and charted what I believe to be the two methods of communication. One inner and one outer path.

Andrew is correct I have crossed into Shaminism at this last experiment. No instrument we presently have in science can detect this scalar interaction so it becomes "experience verifiable" only. I could work on a method to detect opposing torsion fields.

Don has pointed out to me that "information" may be shared freely like I have been doing. However device design may be patented, and or sold separately, for those who do not wish to build for themselves. "Concepts" are shared, "devices" are patented, "books" are copywriten. Even though the knowledge may belong to all of us.

I have seen others try this, charging for their materials. I have also seen a backlash of very disappointed buyers. Books are probably safe, but I'm not ready for that.

The truth is, in a machine that works by some method we can not understand, there is no way to guarantee results.

If I could write a "torsion formula" to relate all these observations in both realities [physical 3rd density, 4rth density, and mental plane] this would put math to the concepts and might achieve validation.

If Don and Kosol want to patent their crystal ball device, that only stops people from building and selling it commercially.

There is no way we could patent wrapping a scalar coil around a crystal and moving it's consciousness into your mind.

Book sales, for me is a premature idea because, I do not yet possess a device which works for either AG or ZPE.

The inner planes are well documented already. What is a new idea is documenting other densities. I have in fact run into lists of all the other aliens that supposedly exist in our galaxy online. I have no clue if any of this is real or unreal, other then what have seen myself.

The two Alien forms I have encountered, other then the "crystal conscious mental plane plurality storage medium" are what appear to be:

1 - totally human luminous or glowing forms, and
2 - what looks like the Greys.

The Rain Maker Device

The human shaped 4rth density beings have skin that glows, and cloths that are dark and cover their ears. Faces and hands glow. This is from my perspective in 3rd density and makes sense. I do not believe they appear to glow to one another. I have seen them through "overlapped physical space" while they were using a platonic device resembling the 3SD. This communication is not on the inner or dream layers of consciousness, but directly from the waking state of consciousness. This was the most shocking connection because, seeing is believing. It was a "physical overlap". I now believe this is possible. I have written a model to explain it, which totally seems to agree with Wilbert Smiths density model.

The Greys, I have seen on the "inner" planes, and this is different. It is non physical, meditative, and uses focus to see. I sensed them connecting with me, I did not seek them out. After the realization of this connection, I am aware of their presence working on the inner planes, in a very timeless manner.

As these connections of information flow are almost unfathomable, I have gone underground at this point. I was alone during the "space overlap" communication so could not say if others would have observed the interaction.

This might be another method of verification. Make a connection with the 4 D aliens with a group present, so several can observe and record. If only one can see this, then it is in fact in our heads.

If all see the overlapped space, then it is density travel.

The only other option I see is to go ahead and build some devices until we get one or two to work. My first thought was to shoot for a "space overlap communication" between two devices in this density.

A visual connection only. If this form of "space overlap communication" is possible and it is a platonic device. The platonic device I saw them using appeared to glow at the stationary vortex points of the device.

I have "channeled" through the crystal intelligence of the mental plane connection, "while under" the needed shape for this 3 Density device. That of a star tetrahedron. I have documented the correct magnet patterns in the files section. Yet I still am unsure how to build this damn thing.

Two overlapping torsion forces, and only two.

8 vortexes, 4 spin each direction as interlocked pyramids.

We need no more then the star tetrahedron to prove this functionality on this density.

I believe that the chart shows what Wilbert Smiths group was doing, and had downloaded as well, and why he stopped the other beings from visiting earlier on. A "comprehension" of what is going on here brings one to a position outside "power and control" over others. The two cannot be compatible. My attitude of free expression for everyone and to see us all as equals is a requirement.

The crystal device if completed will allow others to experience this "comprehension". The coil is the key as Kosol has been saying, but not to tap power, it is to invoke an alternate torsion force complete within itself to allow detachment from the earth vibration.

The sliding I experienced was only in my mind as near as I can tell. To teleport it will become totally physical as well and the danger of disappearing becomes very real, until one is stabilized back into 3 Density for a time.

I will try to diagram or photograph the coil crystal I built yesterday morning, if anyone else would like to experience this kind of insanity.

As to Wilbert Smiths group, I believe I barely touched on what they were probably doing, however we may never know. The scalar coil is definitely more then I expected as a pure scientific experiment, and the addition of a crystal adds an immediate spiritual depth in that it did amplify the crystals normal effects. This was the direct experience I needed to convince me that there is much substance to the design you are putting together here. I would recommend caution if others are jumping into this blind. I am a meditator of at least 20 years and found the experience disturbing at first. I would compare it to Sedona's

The Rain Maker Device

Cathedral Rock or inflow which can bring up negative emotions that have yet to be cleared in ones emotional aura.

I feel the experience I had was one of "mind" or pure mental plane functionality. The chart that I created represents the "realization" of the inner and outer conscious worlds for me, and the integrating of the Spiritual and Physical worlds. Both my scientific and my spiritual self drawing closer together in a true comprehension.

Kosol never ceases to surprise me at each step of my learning process, as I struggle to "complicate" and "quantify" his very solid and Spiritual concepts of form and consciousness. Just now realizing that from the start, about a year ago now, his statements are beginning to sink in. I find myself reading and rereading what appear to be simple statements, and yet contain deeper levels. An example:

> Pi is the circle, the pattern of body or physical form, and phi is the pattern of consciousness.

If anyone is looking to begin a list of Kosol's wisdom statements, or to begin a math model this is the beginning. One I resisted for far too long because of the small attention these relationships get in today's mathematics. If this simple relationship is true, and I state this as a scientist, then it may be the most remarkable comprehension of this age.

I am now very much interested in the dodecahedran form as well. This is the first platonic form we studied, the first one introduced by Kosol as I recall. I studied it inside and out, I recorded the formulas for diameter and sides. I drew drawings and I mapped it on basketballs and other spheres, and tried to conceptualize ways of putting it to spinning objects. These mental exercises have left a pattern in my mind, I now even dream of this form. I have yet to begin to work with it in a device personally. I have spent much time working with pieces of the concept, and look forwards to a completed platonic model.

I must say this project contains three elements that I relate to very strongly at present. Pezio crystal, scalar coil, and platonic form.

If there is a step by step method necessary to comprehend what is happening here I would recommend as beginning steps:

1 - Obtain some quarts crystals and begin to learn to vibrate them with your hand chakras. This can be felt as heat, vibrations, tickles, pressure, or crawling sensations. Sensitivity to the pezio crystal structures is necessary in my opinion. Different crystals do "feel" different. Develop a sense of this reality.

2 - Take this to the next step and move consciously into a crystal, mentally expanding it around yourself as you hold it in one hand.

These are both common meditational techniques found in many crystal books available. But I really feel this sensitivity will aid in learning to interface with this sort of device, first at the lower chakras or hands and finally at the higher centers of the mind.

Next I recommend picking up a stack of Neo magnets and do some of Kosol's third eye meditations. Work up to stronger magnets at the third eye center, but start with only a few at first. The North pole I have found has an ability to bring in a light releasing of tensions that may form during a scalar type or torsion meditation experience. They restore the natural 3 Density balance of mind to this plane. In Sedona I would go to Bell rock for this effect after an intense inner processing of my negative suppressed trauma events. The process of auric cleansing or "healing" which is becoming more prevalent today. A disturbed "emotional" body is going to get in the way as we open to this type of communications. And if indeed the "wishing stone" concept is realized, only a person who loves himself and loves others is likely to do well with it.

When one is comfortable with these things then they may naturally begin to understand the concepts of this project and the possible benefits to humankind.

The scalar coil may be a total mystery to some, and I have studied magnetism and overlapping magnetic fields every way that can be imagined. Overlapping magnetic fields form a force "paradox." A "paradox" is one of those things that the great

The Rain Maker Device

meditators of old used to awaken their students. We can form mental models of this but they all seem to fall short where two opposing fields actually equal exactly. However "experiencing" the torsion forces has clarified this tremendously. The experience was however "familiar" in a way I would never have expected. The vortexes at Sedona which I could not measure with my compass and yet I could strongly sense was this familiarity.

I look forward to following this project closely and I thank you all for the honor to participate.

For the highest and greatest good of all involved. The coil structure is simple.

About 12 feet of 24 gauge wire wrapped down a coil form such that each side of the wire moves back and forth down each side of the coil. The wire does not wrap the coil 360 degrees only 180 degrees.

Start at the center of the wire wind a loop around the coil. When you get to the opposite side of the first wrap, twist the wires around each other so that each wire reverses direction back the way it came. Do this on both sides so twists run evenly down two sides opposite one another.

The idea is to create a coil where each loop has Electron current in opposition to the next winding.

As current is passed through the coil, the Aluminums Protons are turned into the next layer next to it. This creates an opposing Proton magnetic field.

This creates a time vector that begins to extend the Proton magnetic field out perpendicular to the coil all around its sides.

This field contains a time compression in two of the 3D time vectors of the field. The third vector of the 3D space expands outwards allowing the field to couple to the mind. The Proton magnetism actually expands outwards a great distance, several feet off the sides of the coil. This is a compressing field that pulls inwards on itself.

The Electrons magnetic fields fall into a state of looping parallel motion down the coil and do not extend outwards. Conventional electronics would describe them as cancelling one another, however since the "magnetic" flows around the wires are in parallel they simply contract and pull inwards.

This coil structure shrinks the Electron magnetic field and extends the Proton magnetic flow outwards well beyond the Electrons field.

This is an example of the Protons magnetism extending outwards in one dimension. Yet it should still contain the "cold" inwards contraction of the other two directions of the Protons magnetic field.

We are accessing a magnetic field that little is currently known about. The Protons magnetism is said to be cold or sucking inwards on itself. The key to extending it is in the tripole magnetic structure, where at the face of the two opposing magnetic forces the field shoots outwards at 90 degrees extending several feet.

Dave

Detailed explanations from Dave:

For how long do we energize the coil? For a few minutes or for as long as we want to interact with it?

It takes me about 60 sec to begin to feel it strongly after power up. Once linked then I feel the frequency changes immediately.

Can you still interact with the coil after interrupting power to it?

Interesting question, the Aluminum tube releases pretty quickly, but the effect can linger inside me if the focus is kept on it consciously in me not on the coil. Holding the feeling seems to be possible as with meditation.

With the crystal however, once I linked to it, and allowed it deep inside, it stayed for two days, until I used the magnets. The coil was powered for only a few minutes maybe 15 to 30 or so after I linked with it mentally. My time scale was altered however. This also could be my own lack of ability to remove my focus from it. As with the Kundalini type of energy, it sometimes has its own mind and knows where to go and what to do. Some have had it

The Rain Maker Device

remain stuck on for weeks. This is why I described the effect of the crystal as a "conscious" effect, and the description I "went under" denotes a greater mind was manifesting control of the interaction. Sometimes the Kundalini must be "ridden out" till it is done, until one learns to trust and not to resist its force.

There is a way you can check whether the wave from your coil is EM or Scalar. Place the compass in a Faraday Cage, if there is any interaction, then you will know for sure that the wave is scalar since EM wave can't penetrate the Cage.

I assume you mean "scalar cancelling."

I would expect only an Aluminum or Bismuth cage would affect it, but this could be interesting if you care to try it. Since the 90 degree field turns a compass it is not "scalar cancelling", yet the result of a scalar cancelling interaction inside the Aluminum atoms. I describe the field as a Proton field, however it may be a combination of bent Electron flux and Proton diamagnetic flux moving through it and extending out further.

Try it yourself and see what you get, then we can compare notes.

See if you can perceive the energy.

I will post some more of my present models in the files section in an attempt to share my current feelings on magnetism.

The word "scalar" has taken on a "magical" meaning. This means that we do not understand it but recognize the importance of addressing the fact that a unique circumstance is present we can not explain within it.

In fact the word scalar only means having a "force" with a finite value of amplitude. When we add a "vector" to the "scalar" it then has a direction and can interact with other scalar vectors to create resultant forces. The "scalar coil" is really a slang, and indicates a coil wound with two fields that produce strong attraction to one another. One has North up and one has North down and they overlay one another.

As with atoms when magnetic fields are setting in attraction they do not project outwards as far, but pull inwards shrinking their donut size magnetic fields. In traditional electronic teachings this is cancelling the fields. However it is apparent from the study of models like the Sweet VTA there must be some kind of scalar vector interacting at 90 degrees to the main ones as they contract inwards.

I have broken the model of any one magnetic field down into 3 scalar vectors of "time flow rate" in 3 dimensions that are relative only to the line of flux originating within the atoms motion that produces it. Combining them we see some interesting results.

The only forces that can have an effect on other matter are the ones that do not balance perfectly. As we cause the two scalar windings to balance perfectly, why is another field appearing at 90 degrees?

You bet this is confusing and unexpected, and this is what makes it so important. We need to study the nature of this field and recognize it is not the same as conventional magnetism. Or we need to disprove it, and show the error of the experiment before another myth evolves! LOL!

With understanding magic becomes technology.

Kosol on pi and phi 5 - 12 - 2006

As for the higher shape for the pyramid there is only the sphere. The pyramid is the connection with the dedecahedron. Notice Dave what I shared with you mathematically, the ratio of pi is 3.14 is for the sphere, and the ratio of 1.86 is the mind and scalar wave and consciousness torsional field. All of this ratio is connected with the platonic all of them including the pyramid as well use both of these two ratio of phi and pi. You and I know that the universal in form us pi and the universe in energy scalar wave use phi. nNw you understand the mind is phi and the body is pi. This is very crucial to the fundamental of the universe. You have been my student for one year. I have nothing to hide from you as you can see I break everything down to its basic principal of pi and phi as the fundamental universal mathematical law that governs everything in the macro and cosmo of creation. All platonic share this two ratio of phi and pi, so all aliens both super advanced and advanced share these two fundamental concepts. As you can see both of my devices share it also the crystalline 3sd (celestial wishing stone) and the mechanical 3sd. As you can see the perfect form of all platonic is the sphere. It is the perfection of phi and pi in one of what we call love vibrations. You will notice that the grey nor even regular human can catch up with me or the guardians unless they understand that pi and phi ratio is love. Is how consciousness operates through forms and scalar wave, that created our physical universe. Consciousness use this particular principal as medium for it to manifest anything it wants this two ratio and the platonic solid.

Best regards,
Kosol Ouch.

Ok, here is the meditation method. Meditation comes from contemplating on an object of visualization.

For example, you can visualize the breath and see it flow with you mind eye, going in and out, feel it, ask the question of what, when, where, how, why, to it. Is it hot, warm? Do you feel it when you breath the air where does it go, etc. Contemplating on the object call breath you will gain an understanding of it. You continue the contemplating with the why, what, when, where and how question. You continue to follow the breath, where do it go, and you do it until you enter a trance like stage where you no longer are here. But you consciousness has left your body, this procedure of contemplating with the five question can be used with any object of visualization or object in reality with physical visual. Remember you continue to contemplate the object of meditation with these five questions until you enter a trance like state similar to a day dream. The ancients call this samahis or deep trance.

Ok, now you can begin the reading of meditation method below. Bear in mind that these meditation methods use contemplations of what, when, where, how and why questions to follow the object of meditation until the meditator enters a trance like stage.

Subject: Levitation Methods, From the Guardians and Kosol

There are three chakras that are associated with levitation. One is the root chakra or first, second, is the heart chakra or the 4th chakra, and the 3rd is the third eye chakra, known also as the 6th chakra. Just choose one of these chakras and meditate on it. Then you possess its power to Levitate. Remember if you develop one system you are also developing the other at the same time. It is like killing a million birds or more with one stone. For example if you meditate on the sixth chakras, yes, you will posses the power to levitate but also you will posses a light body and clairvoyant, telepathic, and telekinetic power, also. Do you see what I mean? If you meditate on the fourth chakra, yes, you will levitate but you also develop love energy as well. And if you meditate with the first chakra, known also as the base chakra, you will levitate but you will also develop a deep planetary connection with the earth and her beauty creatures, forms, Mountains, and life forms of different sorts on earth, as well as in yourself, deep appreciation, inspiration, and admiration for physical health and life caring.

Well, another system also known for levitation is the udana, or the upwards flow of life energy or prana. The udana is located at the base of the throat, all the way to the top of the head. The udana links the 5th, 6th, and 7th chakras together. The udana is the upper part of the susunna and the susunna is the cord that runs up and down the spine and links all the chakras together (from Chakra one to chakra seven), from earth chakras too human and planet and star chakras. The udana looks like a vertical cord where all the chakras pour their life energy of consciousness and information into it and that's where they all collect. So all you

The Rain Maker Device

have to do is focus on the upper part of the susunna or also called the udana: the reason it is call the udana is its a Sanskrit word for upwards flow of life forces. You see the udana is very important in that it helps the chakras to exist, integrate, and stay connected. The udana is responsible for swallowing, putting a person to sleep, and controlling hunger. It is also responsible for levitation, moistening the body, and helping the immune system, and is also responsible for dimensional travel like teleportation and walking through walls, rock, and other obstacles, etc. So you just visualize a small cord the size of your pinkie (fingers) and made it golden and white blue in color. The visualized cord or udana should be from the base of the throat to the top of the head in the middle of your upper part of your body.

Now just maintain that visualization and breath energy into it every time you breathe; you'll see the udana become more visible in form and in color and along with your visualization become more real and vivid along with every breath. This will give the power of internal vision (better than x-ray vision). Believe me I see women and men all the time even with their cloths on as naked (without cloths), and also their internal organs, life matrix (aura field), and cells or DNA, etc.

Now once you feel like you got the hang of it then you add to your visualization that the size of the udana grow larger and larger until you feel that it is big enough to carry you up wards and rise physically, emotionally, mentally, energetically, and spiritually off the ground, water, obstacle, mountain, and country, etc. Once you become good at levitation you will also, at the same time, develop the phase body shift. This is the true reward of the udana system phase body shift and is another freedom that the light body offers. This freedom allows your body to walk through walls and phase into the earth like it was water. So you can swim in the dirt like water and walk through walls like the light beings or guardians. You can also jump to different dimensions at will, that is the teleportation also (you see what I mean now the phrase killing a million or more birds with one stone). All the psychic method systems are connected. If you develop one system, you are also developing the other at the same time. Now you can have fun flying.

PS: Please fly friendly.

Light and love from Kosol, the alien and human archive.

Levitation Part Two

Guardians: Greeting dear one, let's begin. In order to levitate using this method, you first must take one of the meditation position or posture, horse stance like in tia-chi or a sitting position like a half lotus or full lotus like in hatha-yoga meditation. Secondly, with the tongue touching the roof of your mouth, you breathe in and out of your nose always. Now you begin to close your eye or open them (is a choice), which ever you prefer. Next you begin by visualizing a sun that is the size of a basket ball or larger, the sun color is golden like the morning sun, and is about 4 to 8 feet above your head, and it shines and radiates its golden energy everywhere in all directions just like the real sun in your solar system and other solar systems. Now as you visualized this phantom sun, you begin to see it with your visualization that it begins to shoot laser like fluid radiant energy into your head, your throat and your soul seat area, with your inhalation, and that this radiant fluid like energy is collected at the soul seat area (known also as the higher heart, it looks like a purple blue flame the size of a fist), located between the fifth chakras (throat chakras) and fourth chakras (heart chakras) in the 7th chakras system. P.S. you can make the soul seat look like a golden sun also the size of the fist where you will collect the energy from the phantom sun above your head into that area with your inhalation. Now with your exhalation you begin to make the soul seat rotate clockwise or counter clockwise (is a choice) with the golden light collected energy from the phantom sun above your head making the soul seat looking like a sun also it golden light is radiating from your soul seat area in all direction (this is developing and increasing and collecting inner strength). Now repeat the process, of inhalation and bringing down the energy into your soul seat as well as collecting it there from the visualized phantom sun above your head and with your exhalation you see the sun golden fluid

The Rain Maker Device

like energy being collected and rotated either clock wise or counter clock in the soul seat area and its light is radiated in all direction. As you continue to do this, more and more, the repeated breathing and visualization exercise, your body will become lighter and lighter, the proton, neutron, and electron in your body will begin to flow upward. Then you will float automatically, when this happens, continue to breathe and visualize the about breathing and visualizing habit. So you can maintain your lightness of body, and then you must open your eye, and learn to control the direction of your new developed floating ability. Remember you must learn to direct your body directional movements according to your choice of feeling. Because inner strengths obey your choice of feeling. The feeling is like swimming in water without holding your breath.

Kosol: Thank you Guardians, that is great.

Guardians: May this make your love and light brighter.
Bless be dear one.

Now the Invisibility Concepts.

Methods on Teleportation and Invisibility

Kosol: Selamat jarin everyone, on today's subject we are going to have fun. This fun is called "teleportation and invisibility; the journey". So please put your hands together and give your love and light to the guardians of invisibility and teleportation methods and technology division. So here are the guardians clapping and with a warm welcome.

Guardians: Selamat Jarin. Thank you dearest earth light beings for the warm cheerful welcoming us to your many sacred wonderful realms. On todays subject, dearest, we want to share with all of you 'teleportation and invisibility', how to do it and also how to use it for daily activity on your journey throughout this realm. So you can be any where at any time by your desire and thought. So no more need of an airplane or travel ship. You only use these technologies for pleasure once you know how to teleport and become invisible at will. Dearest, let us journey this path together with fun and also exploring this new technology from us the guardians. We now give this technology to you to be used in your daily journey and for your enjoyment. OK, dear heart we are going to learn new terminology today in this journey.

Molecular structure: meaning a network of cooperated atoms linking together to form a substance (proton, neutron, auric energy, and electron). Light: high vibrating molecular structure thus it become light.

Oneness: a process of experience where you can be everywhere at once (spiritually, physical, emotionally, etc.), without losing any consciousness.

Hyper space: a dimensional vibration where time and space are considered to be one throughout your experience and interaction, (fifth dimensional reality).

Guardian's light: is a state of communion with other molecular structured light beings where a connection between your molecular structure and the molecular structure of other light beings, that you have relationship to. This experience will be fully experienced on the fifth dimensional existence.

Light language: is a light encoded experience that can be transmitted to other light beings with the use of telepathic communication and shared experience throughout teleportation.

Teleportation: a process of which light or atom can be transported from one point to another through worm hole.

Chakras: energy network grid system or worm hole vortex energy system. That can be used for many purposes. There are human/humanoids to planet and to galaxy chakras system. All is a gateway to other dimensional existence's and to other points of destination throughout the universe and other universe's.

The Rain Maker Device

Worm hole: is chakras or vortex of energy, that can be natural or artificially generated. It is a gate way that can be used to transport energy of information or matter of any sort from one point to another at will.

Worm hole network system: is a system of natural or artificial wormhole linking together to form a network grid that can be used for any purpose.

Aura: is the bioplasmic energy around an object and life form. It contains divine light, atom, proton, electron, neutron, antiproton, photon, anti-electron, antineutron, and also anti-energy and antimatter particles that exist in and make up your aura or auric field.

Materialization: a process of which energy (is aura or auric fields) is given form through visualization's (is substance or molecular structure) and thus becomes matter.

Color: a process of identifying vibrational energy light expression, on different levels of manifestation.

Guardians: Dearest, let's begin how this technology of teleportation and invisibility works.

Visualization: a process of CO-creating a artificial reality, within the already creative cosmic/micro reality, through the process of thought and feeling, expression and creativity. This process is used to generate artificial or photonic worm holes that can be used for travel and materialization of energy or matter to and from any given reality. OK, here we go. Teleportation 'the practice'. The first order of things that you must understand is that teleportation is a natural way of travel for all beings of light on any realm. Whether it be on the third dimensional or fourth, fifth, and higher dimensional existence. It is natural and very easy for any light beings because it all deals with "light." Different colors of light give you different level of awareness. Thus your experience will be varied. Also throughout each level of light expression. OK, now since we have a common understanding of our vibration here let's begin the method. Visualization is the key. This will help you navigate and go to any place you want throughout the portal.

One: Just pretend where you want to go and thus see your self in that place already, just look at your present environment, of where ever you are at. Now pretend that you are light (can be any color of light form) and that you can be any where on any given time just like the sun light that radiates all over the solar system. Once you have that kind of attitude developed, then you can see that your destination environment has materialized to you or you materialized to it, through an artificial generated portal or worm hole. This portal can be big or small, that is big as a house or small about six to 10 feet in diameter or so, as well the worm hole that you have just created can be 6 to 8 feet in front of you or close to you. This portal is just a pretending one also created by your visualized thought process. You pretend to create a portal or worm hole with your thought or visualization skills. Thus you give this artificial wormhole color and form like black, white, or blue color, etc. Then you see through this portal (in a form of that chosen color vortex) the desired destination that you want to go. Is on the other side of the artificial created portal. Like a two way mirror except this one you can walk through and be on the other side of the desired destination which is on the other side of the visualized portal. But make a real affirmation that once you go through this visualized portal that you have dramatize through the pretending process. You will be transported to the desired destination, to that place you see, the desired scenery, the tree, mountain, water, rock etc. see it, now remember it is still pretending. Thus see yourself going through this portal with your mind's eye and visualize yourself on the other side and pretend that you are actually there. See this with your mind's eye and with your physical eye open or closed. Now once you do that you will find yourself actually there physically, spiritually, and feelingly as well. Congratulations, you have teleported to the desire place of your destination. That's it. Also you can physically walk through

The Rain Maker Device

the artificially created visualized worm hole, while you are still pretending through your visualization. Thus the affect is still the same. You will be teleported there physically to your desired location of choice to anywhere in the universe or to any dimensional existence.

OK, let's review: One, we know you are presently in your current environment.

Two: We know you have created the portal to your desired destination which is on the other side of your phantom/photonic generated worm hole with it's desired color etc.

Three: We know that we have to pretend, that once we walk through this visualized wormhole or vortex, that it will fold time and space for a brief moment and transport our molecular structure and light form to the other side which is to our desired destination and place that we want to go.

Four: Now once you visualized the wormhole, in reality, you actually created a real one, and since it is close to you or your form, that is 8 feet or closer to you, this will affect your molecular structure and your own chakras system. You at that point are raised in vibration and also are raised in dimensional experience. Thus you are not in a third dimensional existence any more you have actually been, for that moment, you have been transported in molecular structure, light, atom, experience etc. to the fifth dimensional existence, in that fifth dimensional existence during that close encounter with your artificial created behavior the same way that it warps/fold time, space, light, molecular structure etc. Thus this will take you into the fifth dimensional existence through the process of generating this artificial worm hole with your visualization for that moment, from this point of experience you can go to any point in the universal order whether it be time or space on any living event or event that you haven't experienced yet or people from your timeline. Yes time travel is a reality as well also as dimensional travel too, or any universe, no limit what so ever.

Remember dear hearts, that teleportation is not just only for transverse of time and space, it is also allowed you to cross dimensional existence's. You can teleport from third dimensional to fourth or to fifth dimensional or higher existences. This technology has infinite possibilities.

Kosol: Now everyone, the Guardians will show you invisibility.

Guardians: Yes dear heart, we are now going to talk about invisibility. The nature of invisibility is from "blocking out light and it's many spectrums", like a cloud blocking out sun light on your world. When light is blocked out things become invisible or unseen. The same way the key to invisibility is the cloud's, but we are going to use your auric cloud dearest. That is this auric or photonic clouds that is around your body or object. Now again, visualization, feeling, and color, come into play.

One: Color that is very good for invisibility is black and the other is indigo dark blue, yes dear heart, black is a color as well. This vibration can counter act visible light, for black is the "INVISIBLE LIGHT". What we are going to do is pretend again. Just see that you have aura around you and feel its presence heavy or light. Now see your aura change in color from its known whatever natural color it is currently radiating and change that natural color into "BLACK". Now your aura is black in color, or make it indigo black, that is OK to and feel that in your mind and form that you have done it, see it with your visualization skill. Dear heart there is no wrong way, your black aura photonic artificial created clouds can be close to your physical form or twelve feet away from your form. It can be thick or thin, as long it envelopes your entire form with it black/indigo color, remember dear heart your aura is still that same color, but this black aura is actually a phantom or photonic generated artificial created aura or energy shield fields that you visually created that is designed to block out light, and harmful matter and energy that you don't desire to experience. So once light is block out, thus you become invisible, peekaboo we don't see you dear heart. Now pretend that

The Rain Maker Device

you can't be seen, that light can't find you, so you have been shielded from the light, so no light is radiating from you either, or the environmental light is now being blocked out by your shield and thus there is no light reflection from the environment to you and you to it, so you become invisible through your artificial created auric phontonic shield. Now you see everything but everything can't see you, there is no light reflecting from you, it is blocked. It is like, again dear heart, a TWO WAY MIRROR, YOU CAN SEE THE OTHER PERSON AND THEIR ENVIRONMENT BUT THEY CAN'T SEE YOU. Now the application of this cloaking technology dear heart is many. It goes even further then invisibility, with your newly created invisible auric shield. You can also stop bullet, and bad energy as well. This shield can be use to levitate or shield you or your friend from any harmful reality that normally you or your friend's would end up dead or badly injured in the experience. Yes dear heart the shield can be used to stop bullet from a gun or a rocket from a rocket launcher. Since the shield is made from photonic energy of your thought and auric field it affects matter and energy around it. Guns, computer, electronic gadgetry or electrical based powered technology will not work around you, once you activated your shield, even your car won't start. For example dear heart if someone shoots you, your shield can freeze automatically the bullet in mid air because the photonic cloud affects both energy and matter and will obey your desired wish and thought. As well dear heart, the photonic cloud generated by your visualized thought process, can materialize any object of your desire, even another human beings, angels, food, technology, gold, metal, animals, etc. This is what we call materialization technology. The phontonic cloud is all around you, it is your aura. So you see dear heart, you are a divine CO-creator of the many infinite universes and also it's many infinite wonderful spiritual and physical creations. You have no limit what so ever dearest just like us the guardians force also. So you can direct your artificial photonic auric cloud to cloak a house or a particular environment, even angel or other advance being even with auric vision can't see it only if you want them too. So dearest this cloaking technology has many useful path to enrich your life.

We now give this to you also. Use it well. Also don't forget your cloud is alive and you have to build a relationship with it (talk to it with love, feeling, compassionate energy, and light). For it will become your best partner during situations, so be friendly to it and give it a name and treat it with respect. You are Batman dearest and your cloud is Robin, the dynamic duo. Your cloud is very conscious and aware, so see it as equal to you and also it is an angel to you. Also remember it is your family. Love it and respect its existence so that it will support you and your safety, for the cloud will save your life or the life of your loved one in situations that you yourself wouldn't normally get out of. Don't forget your cloud can change shape as well as color, you can illuminate your environment with light that everyone (like a light bulb) can see, by changing your cloud color from black into the visible light spectrum, red, orange, yellow blue, green, white, etc., any color beside black or dark indigo. Dearest that means you can brain wash people that you are from other worlds and that you are angel, as well, make it up as you go along dearest, be creative with your new technology super power skill and gift, again dearest. Remember the golden rule for all light beings throughout the many infinite universal order that is "DO NO HARM TO YOUR SELF AND DO NO HARM TO OTHERS". We now take our leave dear heart Selamat Jarin, light being. Be one and be in joy for together we are victorious.

Kosol: Dear heart one more event.

Guardians: Dearest all these methods are very advanced. These methods are also shared in pantajaly yoga sutras from your worlds. By becoming the object of the sun you can shift your molecular structure or awareness consciousness throughout the universe's just like the sun radiates it light throughout the universe's. That means you can do teleportation also by pantajaly methods. All you have to do dearest, is to look at a object with light radiating from it, like the morning sun light or the evening sun light or a light reflected crystal (light reflected from a chandelier). Just look at it with your mind's eye (eye close) or with your physical eye, in time you will merge automatically with the object molecular structure with yours. Thus you dear heart will process its super power, because your mind will see itself to be the object that it observes, then your molecular structure are given the freedom automatically to experience that object reality from its' point of view and yours together and super power of that object will be activated in you to be used at will. By becoming the object of your observation you will posses its super power or sidhis. Also, dearest, you may notice that we speak of the word's super cult. These words are only meant to help the humor in the expression. Because on your world everyone belong to a group. In the past the bad groups are called a cult, the good groups are called the Christians, or Red Cross, lawyer, doctor, monk, yogis, Taoist, Buddhist, etc. But dear heart from our view we consider all of you are a cult/family/groups/ tribe/organization of some sort. Is to be one and the same from our point of view because to us you all are a collective of individual multidimensional talented souls of different galactic and universal origin coming together for a common purpose. That is, to grow with each other in both spiritual and physical evolution toward the divine light. For all of you belong to the divine mother/father god creative force. Selamat ja dear heart, light be one and be in joy's for together we are victorious.

PS: What made them bad or good is the intention and activity that is behind its goal. Free will creativity has been given to help and support the natural order of the many divine universes.

Thanks in advance,
Kosol and the Guardians.

How to Grant Wishes

There are many ways that a person or group of people can grant wishes.
    In other words, this ability is called the ability to materialize desire at will into physical reality from the spiritual dimensional reality.

The first order of things is to create a 'visual environment'. You have to visualize that you are surrounded by the guide's, known also as guardians angels, or some will call these light beings higher humans. Then with that in your mind's eye or in your visualized environment, all you have to do is to tell the visualized guide's what your desire is with audio, mental words or a visualized thought within the visualized environment. So that what you want will come true into this physical reality. During this interaction with the guide's you must feel compassion. For feeling energy made this interaction with your visualized guardian become very real in all reality. As well your request and desire that is expressed toward the visualized guardian will be materialized. When you have this feeling emotion of compassion, make sure that your guardian has a name tag and having an aura field with multi-color or with your favorite color of your choice as well in the visualized environment.

You can do this for yourself or doing this for someone else that you want to share your kindness or wish materialization with. After your request, please be sure to thank the visualized guide's or visualized guardians group. As well you must continue to feel compassionate feeling and know in your heart of heart that it will come true, by opening your heart and mind with no condition what so ever. Remember hate feeling/emotion can kill while love feeling or emotion can heal and make wishes come true.

Also for those who do healing on people or group of people, you can also let the patient or the person/group of people who is currently receiving the healing energy within the healing session, also request the guide's or their group guardian angel, silently or out loud, for their desire to be materialized also on their own during the healing session. The reason this is possible is because during the healing session the healer and the patient's auric vibration is raised to the spiritual reality multi-dimensional level of existence (3rd, 4th, 5th, 6th, etc. dimensional existence). So their higher self is reached and they are one with the universal consciousness at this level. Now they can materialize their many desires into reality very quickly by expressing their desire to the guides, guardians angel or mother/father God force. Don't forget to thank your Guide's or guardians angels afterward and also do not forget to thank mother/father God.

Thank-you,

Kosol and the Guardians

Resurrection and Materialization

Kosol: Guardians, is it possible to bring dead light beings back to life if they already been dead for both short time or long period of time? Because on this star system and planet they're people that die.

Guardians: Dearest Kosol, all, is possible. We are your guardians and your galactic federation buddies. Yes Kosol it is a reality to bring people back from the dead. Here is the method. Dearest to bring dead people back is very simple and is very fun for death is not of the natural order. Death is a disease that have plagued the light beings of earth since the fall of Atlantis (the Atlanteans did bad genetic engineering on its fellow light beings and mutated them from a twelve helix DNA/RNA into a two helix DNA/RNA thus, death was the result they wished to created. A slave human race, hehehe but we put an end to their dark genetic project 13,000 earth years ago when we the guardians force crashed their party of slavery). Now dear heart in order to bring people back you must know the technology of cloning.

Cloning can be done in many forms. You can infuse your thought energy from your aura's cloud to the DNA of the expired light being or loved one and their form will be regenerated and restored back to full. When your auric fields is infused into the body, the DNA will be activated, thus the person brought back. They will contain all the memory of all life time in the present, past, and future. Thus their DNA/RNA sequence will be a twelve helix light strand with light body. As you all know, dearest, death is a disease and the cure is light/love/compassionate technological super power. Light beings were never meant to die. Death is a new thing on this planet. Human's are meant, as always, to live forever on any realm as they choose and after for any time, by their free will, they can ascend with their physical form to the higher realm by their desires and their angel will meet them there on the higher dimensional existence. So here we go, dearest, go and take a dead light beings. Can be fragment, a whole dead body or even you. You don't have to leave your room, just teleport one into your room. Once you obtain the desired dead human light beings now ask all of your guides to stand aside, so they won't get hurt during the energy restoration, to clone the person back to life.

Using both the methods of invisibility and teleportation you will levitate the dead body with your invisible cloud's (make the color of the cloud is golden) and materialize a crystal bed from the astral realm with your auric golden cloud's. Now with the fragment of the dead person or the actual whole dead person. OK put them on the crystal bed that you have just materialized. Now stand back twelve feet, again dear heart use your cloud's to form a teleportation vortex 2 feet over the dead person's body or his/her bone or hair fragment. Once the vortex is formed you need have to affirm that your guardians will summon the other guardians from the ninth dimensional existence to pour out multidimensional or inter-multidimensional regenerated energy/ light through the teleportation vortex from the ninth dimensional realm that you have created from your auric clouds. Now with your command let the vortex come down and envelop the dead person. During this state the devas, angelic force of creation, and life energy will infuse the universal multidimensional life force into the fragment or the dead person. So the cellular regeneration of the cell can begin and are animated and the energy point network, chakras system, their light DNA/RNA sequence are being holographically/physically/ energetically restored and regenerated into a light form of the person. Everything that they are, is being made anew. They are beings infused by multidimensional energy, dearest, as well being cloned, simultaneously. About 1 hour later, depending on your concentration, your dead friends will be fully restored to life a new as an angel human with light body.

Then dearest you must telepathically or audibly thank the devas, archangels, guardians, of the different dimensional realm

The Rain Maker Device

for team working with you throughout this cloning reconstruction process. Now seal the vortex, by making the cloud dissipate, reverse its vortex spin from counter clockwise to clockwise, see now the cloud that created the vortex is being dissipated. Now go and greet your new resurrected friend from the dead and enter your new adventure together with your new friends dearest.

Now let us review what has happened.

One: Dear heart you all must have something that was apart of that beings whether it be a hair sample, bone fragment, or the whole body it self. Remember never attempt this kind of technology usage if you have not mastered teleportation or invisibility super power technology because both technologies are needed for this kind of super power creativity activity. Dearest, for even those we of the guardians can do all of this at ease. Even we also on some occasion require the elder guardians to be by our side from time to time to over see our activity in relation to this kind of cloning resurrection process. You need a physical and spiritual angel team network for this kind of advance activity. Although dearest you can do it alone, just you and your super power, but is good to have a team of angels, guardians, devas also. It makes things easier then doing it alone.

Two: You must have a plat form to put the body or fragment on, it can be materialized from the astral realm or can be created on the third dimensional material like stone or woods, but it will cost money thought we noticed your world uses currency. Is a very challenging system of civilization's. Anyway dear heart, you need a flat form, or a flat rock will do it also. It can be rock, wood, metal, anything that is hard, and tough, because dear heart this element will interface with the other dimensional realm when the vortex is activated, thus its molecular property changes during that state. So no soft material can withstand this kind of quantum dimensional interaction.

Three: Now dearest you must place the fragment of the person, or body, hair, etc. on the hard rock /metal platform then you are good to go to the next step.

Four: Now summon the many guardians force and type of the universal order. Like this, you guardians of the many realm, come forth to us. If you dear heart can't see aura yet, then just summon them by requesting them and use you imagination and creativity to see them there. Don't worry if they really are there or not, because dear heart when you call on us we will come, regardless who call us, we will come even if you can't see us, notice we will be there at your request and summons. Now just say guardians, angel, devas, archangels, be with us now in this adventure, you must learn to lie to us in a good way, so we of the light worlds won't get scare of your seriousness, you must dear heart learn to brain wash us, in your favor. We notice on planet earth humans have the ability to brain wash each other with good creativity, to make each other do things happily, dear heart, so you must use the same creative technique on us also, we all believe in everything's that you all say. Your voice always entrances us angels of light. So you must learn to brain wash all of us into your service. We are like children in the high/higher sphere of order, although we already know everything that you going to do, but we love to participate and play dumb, hehhehhehehheheh, like we don't know, but we already know your heart desire dearest hehheehehhe, within your drama. We ask that you make it worth while for us also like Shakespeare in the movie on your worlds adventure dearest have brought us to your sacred shore. When we said lie, we want you to go all out in trying to dramatize us into your game "make the story and play look good, like your movie character WILL SMITH SAYING DEAR HEART, we going to make this creative adventure look good". Now you understand dear heart.

For example dear heart, "angel of light, angel of darkness, angel of the east, west, north, south, and all the angel, devas, archangel, guardians of all realm and element, etc., hear me, I summon you all to my present to assist me in the resurrection of my friend, girlfriend, family member, etc. You all know the score

The Rain Maker Device

dearest now be with us in the moment and dance along with the drama story of the creativity, etc. Just dramatize what you want for us to do and to help you with, that all, make it fun, and silly as possible and we will come quick to your aid. We are attractive to child like energy as we are child like our selves. Don't be 100% serious, just be 20% serious and the rest just humorous fun and excitement.

Five: Now since everything's is in place including us in your midst also, it's time for step five. That is beloved you must summon your cloud's that you have used for invisibility practice and adventure. Do this, computer activate my cloud, just play with it dearest like you are in the star trek holodeck. Once your clouds are activated, now you must immediately change the color of your cloud into golden colors, command your cloud to radiated gold/white light. Now tell or direct your cloud to form a vortex two feet over the dead person lifeless form, bone fragment, hair, etc. Make the vortex about 12 feet in diameter dearest. Make the vortex run counter clockwise, once this teleportational vortex is form and then tell the guardians, archangels, devas, etc., to do their things to bring in the energy of the creational force through the vortex from the nine dimensional existence and transfer the essential multidimensional light energy into the lifeless fragment, dead body, hair fragment etc. This energy dearest will clone your former love one, and restore them back into life and a human being again. The hair or bond fragment will begin to regenerate automatically once this multidimensional light infuse with its DNA/RNA structure from the whole person. As well the dead person will be restored into life, from the multidimensional energy from the ninth dimensional of the guardians and archangels existence through the vortex, as this energy infuses itself with the dead person or their remains dearest, their DNA/RNA will be restored, activated, regenerated, as well the chakras system energy network, also. Beloved their molecular structure is being rearranged into a higher body format that is being transmuted into a light body with 12 helix DNA system network grid fully activated. This is the true human body and mind that is a light body with a 12 helix DNA strand and not with a two helix strand like you all currently have In one hour or less, your friend will be fully and completely restored back to your worlds with an immortal light body of an angels form. Now once everything is complete you must thank the many teams of light beings, guardians, angels, devas, archangel etc. that has come to your aid and mind game to help your beloved so and so. Now you must disintegrate the vortex by reversing the spinning from a counter clock wise direction to a clock wise direction and command the vortex back into your natural auric field from which it came. Now go to say hello to your new angel like friend welcoming them back from the dead, dearest. Let's invite them to get something to eat and drink, we are hungry, dearest make it up as you go along. The point is your friend is back and that's what counts.

Guardians: Now dearest we are going to show you another one which is called materialization. This is easier, all you have to do is open a teleportational vortex the same way that you did with teleportation. Now project your memory of the person you love or lost onto the teleportational vortex screen. Once you see the person in your vortex screen, just reach your hand into your vortex and grab him/her from the other dimensional existence. Once you do that, just pull him/her out into your realm from the other side, now you got them, they are now physical, but be careful they can again also be harm or killed once more you see, they have a third dimensional body again with out incarnating. Hehhhehehhe, now the other methods you can do, is phamtomize, that is you can project your beloved into reality by your materialization super power. Just ask your cloud to take the form of your beloved, and you project your memory of that person into the cloud. Now the beloved beings that you love is back but it is a phantom. These beings will act and behave as your beloved before according to how you know your beloved person when they were on your realm with you, but once it served it's purpose it will disintegrated, unless you give it free will, and allow it to choose to die or live. This type of beings will learn to evolve and grow. They only exist if you want them to, other wise they will return to the astral world's to continue their life over there. Or

The Rain Maker Device

you can ask them to live out their life on earth with or after you ascend etc. Then they will comply with your direct or many desires and command. Also they can make baby with a human, they can get sick, or get scared, but they are the same things as you or me, human, light being, etc. The only difference is, is how they got here or were born. You and I were born of mother and father dearest, but the phantom was materialized into a fully grown adult in one second flat by you or another from other forces. They automatically contain and have with them all the knowledge of everything you and I know without going through the multidimensional school. We, just like you dear heart, have to learn the long way, going to school, but for you earth school, for us, galactic federation of light school, but for the phantom well they already have all the knowledge of experience of the person you have created them after, but they will continue to learn new things more and more once they are materialized. We guardians have to say that in a way, we are jealous, ahhh, got you dear heart, yes we are in this case envy of this kind of technology, but we our selves just like you can create and are the parent of this phantom light being 's. Once the phantom is created it is our equal. It is our son and daughter dearest. The phantom will possess the power also to create and destroy just like all of us as well. They also can at will create other phantoms. They have no limit what so ever just like you and us dearest, so be careful when you bring one of them into your realm. Don't get it mad or it will see you as a threat to it, and it will give you a hard time on planet earth. Dearest you must show it love, and educate it our phantom with goodness, light, love, etc., just like you would with your young brother and sister. Good luck dearest, we now take our leave, so selamat gajun. Again this technology is very advance so dearest always follow the golden rule for all light beings through out the many infinite of universe's that is "do no harm to self and do no harm to another" we now take our leave selamat gajun selamat ja.

Light beings be one and be in joy for together we are victorious.

Kosol: Wow, guardians you have given the humans of this world the secret of resurrection.

Guardians: Yes our beloved Kosol, and is all good, is for the good of all.

Kosol: OK, then that is great, hey everyone, you now have the knowledge of resurrecting people from the dead, and also you can clone them using super power technology, your auric fields.

Have fun everyone.

Universal Light Language

Kosol: Here is a gift from the Universal Federation of Light, "The Guardian Light Language" for all to learn.

Kosol: We want to introduce the Guardians of Light language. She and her many entourage are here to teach you all about universal Light language. Please give her and her team a hand and support here in the star gate forum.

Kosol and Guardians: "Selamat gajun, Selamat ja" (light being, be one and be in joy for together we are victorious). We bring you all news and knowledge of the Light language: first, all Light language is in geometric form and second, they are colored from red, orange, yellow, green, blue, indigo, violet white, gold, and crystalline. For example, red is physical or things that can be touched.

1. Light color "orange" is emotion and fluid things that can be felt but don't "need" to be touched.

2. Light color "yellow" is logic/linear construction or pattern and organization.

3. Light color "green" is nature or the natural order of the universal on all levels.

The Rain Maker Device

4. Light color "blue" means a plan or idea to follow or what you can say is a law and goal to be carried out.

5. Light color "indigo" is what we call creative thought, telepathic exchange, and idea generation and/or regeneration to create or to reform.

6. Light color "violet" relates to knowledge connection and usage.

7. Light color "white" is what terms as clear expression with divine support and guidance as well as healing.

8. Light color "gold" is unification of all that there is.

9. Light color "crystalline" color is what we call an expression of total all for one and one for all.

10. Light color "black" means peace and serenity.

11. Light color "crystal fluid gold" is universal body/mind/consciousness.

12. Light color "crystal solid white" is universal agreement, cooperation, goal, and understanding.

13. Light color "crystal line rainbow color" is all is one and one is all with full unity and individual diversity.

Guardians: Now beloved you all know the alphabet of Light. Now you must know its geometric support.

1. Light symbols "triangle or pyramid" means gateway or connection to a place of or within or to energy and knowledge as well as experience.

2. Light symbols "circle or orbs" means continue or eternal ever changing and never ending.

3. Light symbols "square or four sided cube" means universal or all corners and directions are to be accepted, respected, and acknowledged.

4. Light symbols "rectangle or long four sided cube" means dimensional crossover or travel, or a journey that is being taken on.

5. Light symbols "cords or circular long cords" represents links or universal existence and reality.

6. Light symbols "wave or many wave" represents agreed event of both group and individual choice of creativity and event.

7. Light symbols "matrix" represent interconnected life on this universal and all the universal order.

Guardians: Well beloved, this is "the universal Light language."

Now we will show you how to put this Light language together. This Light language is used extensively among all universal Guardians force and on a personal level to, since we are the caretakers of the universal on all levels and the many dimensional existences. The many "GOD CREATIVE FORCES" (in all the different many universes whom we all serve and represent by being caretaker of the universal or universal angels) gave this universal Light language to us and to many different universal Guardians species so that we can have a common understanding of each other. Also it is to understand our common goal which is to help carry out and fulfill our many universal goals or tasks/habits to be caretaker of this universe and other universes as well. This universal Light language corresponds to the 13 or 14 chakra systems or Light body and galactic body/mind chakra matrix. (12 helix DNA/RNA systems).

Guardians: For example: "red, indigo, orange, square, green" means multidimensional experience of both perception from yourself and others put this Light color and Light symbol through

The Rain Maker Device

your experience and then say what you see and experience through its definitive meaning. This means beloved that "the universal experience which I am expressing", "but in order for me to find it", "I must first find my own feeling and so by doing", "I can have direction." "And then everything will be complete." Beloved think as each phrase is the definition of the symbol but put it in your perceptional view of experience and as well from other experience that is around you.

1. This is you. "The universal experience which I am expressing". This whole phrase is you or I (first person of representation and expression this is physical and is red.)

2. This is your desire. "But in order for me to find it". This phrase represents your desire. (Indigo or creative thought)

3. Light color "orange": "I must first find my own feeling. And so by doing" this phrase represent it. (Your feelings)

4. Light symbol "square": "I can have direction." this phrase represent it. (Direction)

5. Light colors "green": "and then everything will be complete." This phrase represents it. (Universal common ground)

Kosol: "Blue, green, violet, plus triangle" means I have the responsibility to bring into balance the energy of my own chosen path, and thus I will have the energy to see.

Guardians: Kosol very good it follows the format of the Light. Remember this is basic.

Kosol: Continuing on, "green cords, red blue square, white matrix" means my natural content is to be open to all that is around me and within me. (Green cords) with my life I will carry out my life purpose. (Red blue square). With this the Guardians angel and the earth as well as my human brethren will support me and I will support them so all will be complete within the white matrix)

Guardians: Very good. This is how aura language and Light language is read and expressed. It is expressed from your point of experience, or you can express from other life form experience on a single dimensional reality or multidimensional reality. Light language has no limit what so ever. Light language is universal in its multidimensional expression of experience (based on five senses, free will, commonsense, creative imagination, and diversity). With common meaning it brings truth and common understanding to all that use it and express it. This will help activate your current 2 DNA/RNA helix into a 12 DNA/RNA helix also.

Kosol: We will look at red, blue, wave, and cords. When you see red, that means the person is presently experiencing a physical (red) responsibility (blue), and is having that experience shared with others (wave). The cords mean this person is very interactive.

Kosol: Let us clarify more. The reading of light language is automatic; you must see the color as living experience from your point of view or from the view of another. Once you encode the key code or chevron (light color and light symbols encoded) the star gate of understanding will open up to your experience. Then you will flow with the understanding and along with it, wisdom.

See, as the colors are mood or point of views (on a multidimensional level of choice and agreement) that you must speak from. Each view will describe the same event or understanding to you and you have a choice to say it from the person's point, your point, or from a collective point.

For example: With red you must speak from the physical point of experience. With blue you must speak from the responsibility point of experience. With wave you must speak from the point of view of agreement with other life forms. Cords you must speak from the point of view of connectives (or a connecting relationship) and acknowledge that realization.

The Rain Maker Device

Light language is multidimensional; it is designed as a universal language. It takes everything down to the first basic existence to the most complex. Now what will dictate the goal of the conversations will depend what light symbols it is written in. For example: Let's take a look at blue orbs, green triangle, plus cords and rectangles with orange. Now here is the subjects: to people, to place, to things, or to some form.

The blue orbs: to carry out eternal light by wanting to remember an already learned journey. This is what blue orbs, green triangle, plus cords rectangle with orange mean. As you read this detail you will realize what the light language is describing. Remember the light language describes living experiences and it expresses it from many different viewpoints of the same universal event. It is a multidimensional light language of agreed events. With it is commonness/diverseness in harmonies. It represents and expresses all points of reality and with it wisdom comes forth and is realized.

P.S.: Light language contains your universal intention, both on an individual level and a group collective level. It is to help carry out the divine plan and divine purpose that MOTHER /FATHER GOD HAVE INSTALLED IN EVERY ONE OF US, on all creations levels.

PS: Your "many creative imaginations" will be great for aura reading by using the Light language universal system. The Light language is also called the 13 chakras or aural Light language universal systems.

Subject: Telepathic Talk

When you send your telepathic message (known also as t-mails) be sure to tell your guardian (g-mail) at which timeline or date that you want that person or particular groups of people to receive your message, otherwise the message will be sent to every timeline and other realm to find the person.

Also remember the light language related to the platonic universal geometric which is, the square or cube, octahedron, the star tetrahedron, isocahedron, the dedecahedrons, the sphere, and the isoca-dedecahedron etc. They all follow the pi ratio of 3.14 which is their form and the magnetism of consciousness of phi ratio which is 1.86 just like the pyramid structure of the Egypt and Angkor Wat temple of Cambodia.

Note: Positive attracts positive meaning whomever your message is intended for. They will receive it somewhere in time and the person's alternative self and parallel self will receive it to in other universes and timelines, unless you tell the guardians about the date and universe that this particular person is in. In other words, it is destination or indigenous.

Just visualize or imagine the person or groups of person that you want to talk to standing right there in front of you, about three or four feet away from you or even six feet away from you, and from there get closer up and personal, one foot whether with your eyes closed or open, dependent on your visual skill, development, and concentration in the participation of the drama. See them clearly with appropriate clothes on or just make up one and if you don't remember their faces or just simply forgot, well then just visualize a light being or groups of light beings and the surroundings (environment). Can be where ever you are or just visualize a new environment that you like or prefer. Pretend that you are in the holodeck (like in star trek) and the light being can all have the name of that person whom you forgot (like a name tag or T-shirts with that person's name written on it). If you partially remember, the light being can be male or female (then just talk whatever you think or feel, express it all out).

Remember to use all five senses in the creative thought (sights, hearing, smell, feel, touch, and taste) drama and after the conversation listen to their response (the response would be, audio, imaginary, feeling, as well as actual presentation of the event, etc.). Then after your drama, tell your guardian to send the telepathic information, of interaction, to the person and their guardians as well.

Use this method also to talk to the guides. Just use the part that says visualize the being of light surrounding you and the

The Rain Maker Device

choice of environment that you have created from a real place that exist in third dimension right now, or some where in the future or past. You can even use an event that is in the environment from the astral plane (fourth dimension) and the fifth dimensional reality. There no limits, whatsoever, as long you like and enjoy it. Talk and express, then listen, interact with them, that's all, just like you would talk to a person and a group of people and they talk back to you. Except in this case they are guardian angels or heavenly people or sometimes also known as aliens since they are not originally from our dimension.

Love and light from Kosol.

Develop a Health Immune System With Faith and Meditations.

Subject: A Cure for Aids From the Guardians of Councils and Kosol

Kosol: Dear Star Brothers and Sisters, greetings and blessings from the guardian of council. I'm happy to tell you all, that the council has allowed me to give everyone a cure to aids or the nanite parasite that was programmed to interfere with the human immune system.

Cure 1: For you out there who have meditation skill and have the HIV or AIDS virus, then this cure is for you. First of all, you must understand the area that helps the immune system or created it. The first area is the thymus gland (between the fourth and the fifth chakra or the soul seat) in the upper chest area. Visualize a purple flame big as a triple size candle flame radiating in all directions. This gland produces a lot of hormone just like the hyper thymus gland (in the center of the brain). Once a person focus in this area the hormones are increasing and are produced and t-commander cells is created and can channel 12 helix functions that seek and destroy all known and unknown toxins.

Tongue touch the roof of your mouth and plus the applied of Hatha yoga alternative breathing. The other area that helps you also is the hyper thymus gland that is in the center of the brain. The top part of the hyper thymus gland is the pineal gland and the bottom part is the pituitary gland. This gland produces a lot of hormone, especially the (soma) hormone, the hormone that requires to produce a light body or 12 helix DNA body type. Just focus in this area and visualize golden light or ball of golden light the size of a pin pongee ball (radiating in all directions). Do the same thing again. Tongue touch the roof of your mouth and also apply Hatha yoga alternative breathing technique. In about 6 month time your HIV and AIDS virus are destroyed and body, mind, and spirit are in balance again.

Cure 1 & 2: This cure is related to a two or more parties who know how to use inner strength methods or know meditation methods. OK, position one hand to hand or hand to back, that mean one person sits behind the other person and both of his/her hands are at the back of the heart area, his/her back of the person who sits in front of them, (hand to back). The sitting position is a Buddha posture or full lotus or half lotus and hand to hand is the two people facing each other, and putting hand to hand. Once your partner and you take one of these positions, whatever both of you agree on, then whomever the infected person is, you just do nothing while the person who is not infected will do the curing and cleansing of your HIV contamination to the person who does the curing. First of all, warm up your tan tanium in the navel area (the tan tanium is about one and half inches below the navel and is used by the Chinese Tia Chi or Chi Kung practitioners to store inner strength), for 5 minutes, then gather energy by connecting to the earth's electromagnetic fields (alpha wave 13 Hz per second) by visualizing many cords growing from your base or the first chakra to the earth core. Once you've done that, then visualize your crown chakra is growing connective cords to the sun and other star system formations like the big dipper, etc. Now bring golden energy from them star formations that you just connected to with your visualized cords, and at the same time bring red energy from the earths core that you just connected to with your visualized cords, along with your inhaling of breath, with the tongue touching the roof of your mouth. Now collect all the

The Rain Maker Device

multicolored energy and gather it to your soul seat area or your thymus gland area. Once it is collected, now exhale (remember just normal deep breathing) shoot out 80% of the multicolored energy to the other person's thymus gland or soul seat area and repeat this step for about 15 to 30 minute.

See it in your mind's eye that this person thymus gland is glowing and also growing (multicolor). See the energy is altering the DNA of the HIV. Once you see this clearly in your mind, then you got it. Stage two, now tell the person to jump in the water (it can be cool or warm).

Now use your inner strength from the tan tanium, and bring it up to the heart or soul seat area. Now send it full forcibly to the other person's soul seat. Make it radiate twelve feet around him and keep bringing your developed inner strength up. Keep giving it to his thymus glands and see it with your mind's eye, as well as the soul seat radiating multicolored light twelve feet around him/her and you. Yes, both of you will feel real hot (heats from the inner strength projection), the water will boils as well due to the inner strength, then you will see, the toxin or HIV will leave his body and contaminate the water and die because the inner strength will purified him/her, and clean all the blood, body cellular structure, DNA, as well as alter everything about the person's DNA strands. During this stage both of your bodies are in light body operational mode, so both of you are not normal human in this stage. Both of you have entered samahis (another Sanskrit word for expanded state consciousness), a merging of all conscious together from conscious to collective consciousness. Both of you have become individually one with the earth, universe's, personality, collective conscious, and communion, etc. This process will take two hours or one hour to complete. Do it two times a week for 6 months to a year, then the healing is complete.

Cure 2: The dolphin is the key player with them and they will heal you. It is not a one time shot, you must be close to them or about six dolphins that swim around you (their telepathic capability will rearrange your DNA and the HIV DNA and purge the HIV DNA out of your system). They can and will regenerate your body, mind, and spirit. They will help restore you to your natural balance. They see the aura and have internal vision.

Cure 3: The yogi is the key or a person who has developed a light body of 12 DNA/RNA body type: infect them with the HIV or aid virus for their super immune system will create an antibody. Once that happens take their blood and synthesize it and inject it to those who have the HIV or aid virus. That is the cure. (So get out there and capture a mystical yogi and do what you need to do.)

Cure 4: For scientist, since the nano technology is available in this time line: create little nano robots and program the nanite to seek and destroy the aids virus and restore and repair the human immune system.

Well, that is it star brother/sister, please provide feed back, if you can. This knowledge is authorized from the guardian of council. It is a gift of hope. You'll like it and that it will help you to be creative in finding your own way of curing the HIV infected people.

Truth about DNA/RNA

What is DNA?

Guardians: C A T A G A, your DNA. It is a crystalline structure of integrated vibration pattern holders and channels, as well as a vibration projector and receiver on a multidimensional scale of light and colors (hologram matrix). Your DNA is actually light filament encoded. Matter affects all DNA light filament encoded species: humans, aliens, animals, trees, and on all cellular level life forms. Now the difference is the amount of light and colors that each of this DNA is able to carry and hold, channel, store, as well as receive light and color frequencies. This determines the vibrations and the colors, or conscious perception and awareness for that life form on each level. This would determine the evolution of that species or individuals on every level of existence.

The Rain Maker Device

So vibration, geometrical patterns and colors are the keys?

Remember, all species have a group collective agreement with the universe and God Creative Force, to evolve at a certain pace, as well as on an individual level. Also, free will of choice, on both the collective and individual level, are involved here.

So is everything breaking down to sum up the total?

Guardians: Let's look at this from the human's world perception of everyday science. We have 10-20% that is food and environmental vibration; 30-50% is social vibration; 90-100% is free will vibration, and 100% an infinite percentage, is spiritual vibrations, collectives, and individual free will vibration on the spiritual level. Now the main point here is that all of this is connected to each other, but yet, independent to each other also (in other words, multidimensional in existence).

Is DNA important to super power development and what about the so called junk DNA/RNA that our earth scientists have discovered? Is that important to our development as a species and individually?

Guardians: 100% plus, it is very important to both the individual scale and the entire human species, as well as your spiritual and light body development. The more you act out compassionate feeling, the more you connect to the earth, other planets, star systems, galaxies, etc. Compassionate feeling will activate your DNA/RNA from 2 HELIX DNA/RNA to 12 helix DNA/RNA.

Now as you may know, the scientists of earth have mapped the DNA of human beings, that is what they think, so to speak. Well, they have discovered the so called junk DNA/RNA OR THE THIRDS STRANDS OF THE HELIX'S. The junk DNA OR THE THIRDS HELIX'S is the savior of the species and the planets as well. We call this junk DNA/RNA the miracle DNA/RNA, or light body DNA/RNA. Once a person acts out compassionate feeling on an individual scale or collective scale, this junk DNA/RNA will activate. Once it is activated, your two helix DNA will become a twelve, which is your light body or what the avatar Jesus called the new creations, a body that doesn't die but lives forever. Compassionate feeling is 13 Hz per second in vibration. It is the alpha wave, in relation to the earth, sun, galaxies, and the universe. The junk DNA is a shape shifter DNA. It responds to compassionate feelings and love. It produces and receives and channels the alpha wave's 13 Hz cycle. It is the electromagnetic field of the earth.

As you see, it is all connected on a multidimensional scale. The compassion, love, and free will, as well as Mother Earth, are truly the keys to activate your junk or miracle DNA/RNA.

Light Body or New Creation Body

Method to Develop the Light Body or New Creation Body.

Subject: Light body developing methods-2 stranded DNA helix human beings transmuting in to a 12 helix strand human beings with light body, and how to do it.

Changing a current two helix DNA person in to a 12 helix DNA/RNA person is a one to two year program, to be completed on a massive scale world wide. Because as you may know that, what a two helix DNA/RNA person conscious mind and body can't hold, so the two helix DNA/RNA person subconscious mind and body' can. If whatever the two helix person subconscious mind and body can't hold then their super conscious mind and body can and so forth. What ever your super conscious mind and body can't hold, your collective conscious can and so forth, but they don't have communion with one another in a conscious level of energy and experience all of the time, (half conscious) due to the two helix DNA/RNA can't hold such information and resonate with that experience on a fully multidimensional conscious level all the time. This is what a two helix DNA/RNA person has to deal with every day. That is why we call the two helix DNA person half conscious, where sometimes it know everything and sometimes it doesn't. Just like sometimes you remember and sometimes don't. Or sometimes awake in spirit, and sometime asleep in spirits and so forth.

For a 12 helix DNA/RNA person we don't experience that: what we experience is what ever the collective mind knows and 'experience' what the conscious mind knows and experiences. What ever the super conscious mind knows and experiences, the conscious mind know and experience also, and whatever the subconscious mind knows and experiences the conscious mind know also.

Do you at least feel and an understanding of what I'm relaying to you here?

The key to get to the 12 helix DNA/RNA body (light body type) is that here it is you have to understand the important relationship of the thymus gland and the hyper thymus gland, as well as soma light body hormone to the light body development.

As you know a two helix person cannot understand a lot of multidimensional energy of experience and information simultaneously on a continuous basis or all the time; so in order for a two helix DNA/RNA person to become a twelve helix DNA/RNA person or light body, as I stated before, you must understand the relation of the thymus and hyper thymus to the soma hormone and the light body developments methods. So here is 'key one': the thymus gland's location in side the body (in the upper chest, but below the throat known sometime also as the higher heart or soul seat) is between your fifth and four chakra in a 2 DNA/RNA helix human body structure. It produces a hormone and creates an advance immune system for the body and mind and as for the hyper thymus gland, its physical location is in the center of your brain. It has a relationship or is connected to the top part of the pineal gland and the bottom part of the pituitary gland. They produce hormone (soma, and soma is a divine hormone or light body hormone) and that hormone (soma) has the power to transmutes the 2 helix DNA strands into the 12 helix DNA/RNA strands (The DNA activation process). The hormone that is needed is call soma or light body hormone and in order to produce this light body hormone (soma for transmutation of the two helix DNA/RNA strand into a twelve helix DNA strands) you must meditate on this two point of focus, the thymus gland and the hyper thymus gland.

OK, here is the methods to having a light body or 12 helix DNA/RNA body or what some people's called a Jesus body, known also by Jesus as a "new creation or glorify body type's". Now let's begin:

The first area you must focus on is the thymus gland area by visualizing a purple blue light: like a candle flame on the upper chest below the throat area and above the heart charkra area in the center of your body. The purple blue flame should be visualized right there and that it is the size of a candle flame once you do it for 20 minutes. Then on the last ten minutes you should expand the purple blue flame of your soul seat through your imagination to about 12 feet around you engulfing everything, during this stage you will feel warmth in that area as well as power full. OK. For one you must do this thirty minutes every day.

The second area of focus is to focus at the point or area between the eye brows where there lies a third eye or sixth charkra. This chakra related to the hyper thymus gland, therefore visualize a little golden light the size of your thumb there for ten minutes.

Then the third area to visualize is in the center of your brain or head. Just visualize a golden sun there and the size of that golden sun is the size of a ping pongee ball and do it for ten minutes. After the ten minutes please visualize the golden sun is growing full size of your head and filling your head and shining outwards for about ten more minutes. Then visualize the golden energy as a golden water spout, springing right out from your head through the crown chakras, like water from the water fountain and go up in the air about two feet or three feet and then with your mind's eye again seeing the golden water coming back down and hit, wetting and soaking your whole body, clothing, aura and organs, skins etc. In this golden energy of water bath and showers will create the light body and bringing back or activate your two helix DNA strand in to a twelve helix DNA (light body type) and all of your 72,000 psychic channels or chakras, both major and minor will be open and fully developed.

So the key is practice, practice, and more practice. OK you will have many psychic side effect from this method, levitation, seeing aura, telepathic communication, your hair will turn white or different colors, and as well as dimensional travels, like teleportation and phasonic with the earth and swim in the dirt like water or walk through any solid object will be normal using this methods as well as having internal vision. This is a method for killing many birds with one stone. As well you will become gods and goddess using this methods of psychic development so class do any of you have question? Please feel free to ask.

There are the 7 chakra system for the regular human being, but the light body human once developed with the above method will give the regular human beings or humanoids to have a 12 to 13 chakra system. There are extras chakras in the light body system. There are 13 chakras in all in the human or humanoids' light body system.

The Methods of Healing Cancers Once Your Faith Is Strong Enough

Subject: How to Heal Cancer From the Guardians and Kosol

Kosol: Ok crew, we're moving onto a new subject. On this month's topic we want to discuss that light can be used to cure cancer and aging.

Guardians: Beloved, all this is the truth about medicine in all worlds', on your worlds and ours also. In many infinite realities, including earth, the knowledge of light is the key and gateway toward all of the infinite cause and resolution of all co-creators and creation, in all known and unknown multi-universes. On your world when light is not fully expressed, it gets trapped. Since this trapped light consciousness is a universe within a universe of itself, like all-light beings and light life forms, naturally it would begin to regenerate, and become a new collective reality, a multi-universe within itself, and also an individual reality. This would only happen if the light in your body gets cut off from other light by normal means or by the natural law of consciousness it will adapt and thus it will become of its own and regenerate. This in your world would be ok, except in organic biogeno life forms (human, animal, plant etc.) this cut off regenerated light energy universe and light life form would be called cancer. It would not correspond to the natural rhythm of the agreed dance with the rest of the other light connecting consciousness in the body and soul. So the new light cut off life form will become a renegade from the other light consciousness' point of view that it would normally experience inside the body of agreed rhythm dance. Thus the cut off light consciousness doesn't know that it's not in synchronization with the rest of its sister/brethren light beings consciousness within the person's body and soul. So the cut off consciousness is dreaming, but it doesn't know that it is dreaming. So biologically this is what you earthlings call cancer. And it is in many dream states. So it runs rampage, and it dreams that it is in the collective order and natural. The cause of such cut off is shock, disruption of bad memory and reaction to a shocking drama of experience that leads to some aspect of the whole person to eject a light consciousness out of its system of natural rhythm agreement dance. Only light, love, and compassionate feeling along with inner-strength will heal the light cutoff consciousness and put or reconnected it back with the whole or the natural agreed on rhythmic dance with the rest of body, soul, earth, etc.

Kosol: So what do you do guardians to help wake up the dreaming light consciousness who is being temporally cut off from other light by its temporal unconscious dream state?

Guardians: Dear heart, just simply turn on the light and the disconnected lighted consciousness will wake up and remember that it was only dreaming. Then it will get right back to the natural dance with all of its brethren in the human body and soul.

The Rain Maker Device

Kosol: I don't understand.

Guardians: Beloved, it is all multicolor/multi light. Just look at a light reflected crystal. Or look at the morning sun for one hour. Or look at multi color light sources that don't harm your eyes. Look at this kind of multicolor light, and it can be any color you choose of desire, then all ill alignments/ailments in your body and soul will reverse itself and become balanced with your natural rhythm of natural order again. Good health as well as immortality is reached, this kind of habit will produce light body also. It is the practice of looking at multi-light /color object.

Kosol: Ah, is cinema? The practice of reflecting an object or many objects by just looking at it. Ok I get it.

Guardians: Correct, Kosol. Cinema. Becoming the object of reflection. Since everyone on all universes is a light being. So on the new up coming galactic civilization on earth, light will be used for everything, it is our food source, and as well as all power sources. And also medicine for light body beings and physical body beings. It is our natural infinite energy/matter resource. It is photonic. The light spectrum and photonic light will be the source of all power sources and for everything within our technology, as well for you also.

Magnetism

[It may be helpful to scan the definitions at the end of this section, as well as consider the references on NMR for a mental picture]

Preface

Magnetism has been studied for centuries and almost from the beginning seemingly wild claims as to its importance have been stated. As I too have now studied magnetism it becomes much more obvious that as the 2nd most powerful force we have discovered, these claims may not be so inaccurate. My studies have led me all the way into quantum physics at the smallest measurable places where this force is detected. My surprise is what Leedskin and many of the others have been pointing towards, that magnetism is actually two forces working in harmony. More specifically four forces. It may very well be that in the splitting of these magnetic flows lies the [**Source**] of free energy.
    Engineering magnetism at the atomic level is the only piece we are missing at present. The needed mathematical relationships must be converted to useful formulas to accomplish this for each device considered.

What is magnetism?

At the roots of magnetism, within the atom, we discover 4 different sources which can result in a magnetic flow external to the atom itself. Magnetism is a force linked to 4 separate forms of spin at the atomic level.

The Rain Maker Device

1 - Proton Particle Spin [42.5781 MHz / Tesla] NMR
2 - Proton Orbital Spin [around 2 MHz / Tesla] NMR
3 - Electron Particle Spin [28.025 GHz / Tesla] ESR [Reversed component from Proton]
4 - Electron Orbital Spin [around 12 GHz / Tesla] ESR

The 4 types of spin are all very different and distinct. In the normal magnetic field all these spins are setting in alignment and create basically one polarized magnetic field, in which the Proton is the dominant force in weight [mass is 1836 times higher]. However the Electron has a magnetic moment around 658.2106881 times the Proton. In a coherent magnetic field with both a Proton and an Electron it should be noted that the electron is spinning the opposite direction of the Proton. Its particle momentum is reversed from its magnetic field. When combined within a Neutron this becomes important, as well as establishing an orbital repulsion.

In the non magnetized atoms these spins are distributed such that none align consistently as they orbit the atom. In the atoms of Iron Cobalt and Nickel these spins come into alignment enough to create a coherent flow. In Iron the Electron shell is magnetic and the Nucleus is not. In Cobalt both are magnetic. In Copper only the Proton shell is magnetized and the Electron shell is not.

As an atomic particle spins, its motion contains two different yet linked spin properties.

Angular momentum [the spin of mass]
Magnetic moment [the magnetic field]

The method in which atomic particles interact is through the charge of the particle, the magnetic field, spin coupling, and the strong force.

Charge

Charge is a voltage potential. It attracts Electrons to Protons and it is completely balanced in Neutrons. An inverse distance squared force, reaches further out then Magnetism. Charge causes attraction or repulsion, and the result of a complete attraction creates a Neutron from a Proton [plus charge] and an Electron [minus charge] merging.

Magnetism

Magnetism is an inverse distance cubed force. It is bipolar in any one magnet having a North and South. When Electron and Proton combine to form a Neutron the Electrons magnetism wins slightly, and the Neutron still has a very small negative magnetic moment. The atoms particles all have a magnetic field, as they move about they wobble or precess at some frequency. The Electron, Proton, and Neutron can be thought of as little spinning magnets, spinning in a dual cone shaped pattern, that may spread wider or narrower depending on the frequency and magnetic field they sit inside.

Spin Coupling

This is a hard force to conceptualize, but can be easily grasped if holding a magnet next to a fast spinning copper cylinder. The spin of mass is not coupled by a charge, spin is coupled through a magnetic field crossing Electrons and Protons. In the spinning copper cylinder experiment, it does not matter which direction the magnets poles are pointing, as soon as its magnetic field crosses through the spinning copper the force is felt in the magnet. The two forces felt are "Spin Coupling" and "Magnetism" acting between the Copper and the Magnet. The Copper is a "Proton magnet," and the magnet is an "Electron magnet." The experiment shows why Electrons do not crash into the Nucleus of atoms. As you bring the magnet closer to the spinning Copper you feel the reverse magnetic force that results from the spin coupling. It pushes the magnet away, and also tries to drag it along at a distance, as well as spin it backwards as it orbits. The only thing missing in this experiment is the attracting voltage that would cause the magnet to begin orbiting the spinning cylinder like an Electron orbiting the Nucleus. If released the magnet will be shot off at high speed and hit the wall. So we see that as Electrons orbit the Nucleus it does not matter which way their magnetic poles turn they are still repelled by magnetism.

Strong Force

This is the force operating inside the Nucleus area of the atom which effects gravity. It is 137 times stronger then the Electromagnetic forces, and manifests a strong attraction or sucking inwards of the Nuclear particles. The periodic table shows us that all the parts within the nucleus weigh more when removed [compare atomic weight to Proton Neutron Electron counts]. The Nucleus is converting mass to energy.

These are the forces that we have to work with for engineering AG and ZPE devices. Mainly the first three can be used in device designs but it is probably the last one that we need to reach into for a mass into energy effect.

Angular momentum [spin of mass] is linked to the mass or weight of a particle and determines the particles power to affect other matter. However it is only through the EM or magnetic field that this energy can be transferred beyond the Proton layer. The magnetic field is the agent by which the momentum is transferred. Each of the particles involved has a "charge," as well as a "magnetic precession of spin." **The Proton has 1836 times the mass of the electron, but exactly the same size charge**, only positive rather than negative.

Proton charge $e = 1.602 \times 10^{-19}$ coulombs
Electron charge $-e = -1.602 \times 10^{-19}$ coulombs

The motion of [particle] spin is how the charge becomes two of the magnetic fields.

Protons magnetic dipoles precess around 42.5781 MHz / Tesla
Electrons magnetic dipoles precess around 28.025 GHz / Tesla

The motion of [Orbital] spin is the other method. Magnetic dipole precessional frequency ranges were mentioned previously for the other two fields.

As a Particle Spins it Couples Energy in Three EM Methods

Its [**electrical charge**] is seen to reach out as an inverse distance squared force. This force is divergent in all directions.

Its [**magnetic field**] is seen to reach out as an inverse distance cubed force. This is a raw force of dipole magnetism independent of frequency of precession. It is caused by the electron traversing a circle at .999999..., the speed of light, and spinning a reverse direction, as it does. The Proton doing similar yet with a like direction of spin as orbit.

Its [**RF field**] is seen to absorb and radiate photons at its NMR or ESR frequency as a little point source of RF or microwave energy. This is a link between magnetisms dipole [precession frequency] and the photons frequency.

This interaction has the capacity to flip the dipole completely over increasing its energy by placing the magnetic moment in opposition, and the angular momentum in an aiding state with all opposite particles in the field. The only way that both magnetic polarized states could be stable is if there are a combination of 4 forces, 1 is always in opposition to the other 3. The forces at work here are magnetic polarity of the Proton in its orbital, magnetic polarity of the Electron in its orbital, Proton angular momentum, and Electron angular momentum.

Because the RF frequency is a function of the surrounding magnetic field strength, the particles may interact across a wide band of RF EM spectrum. That of electrons, reaching far in excess of there natural microwave range as well as far below, if they are removed from the atom as with an electric arc.

The Important Differences

The Proton is seen as a magnetic dipole spinning 1 MHz to 400 MHz. Protons are seen as heavy or massive particles compared with Electrons. The Protons orbital motions are seen spewing out MHz frequencies depending on the atoms they are inside which lay ~1000 times lower in frequency then the Electrons. The Protons g-Factor is seen at 5.5, meaning when Protons orbital and particle motion are combined the resulting magnetic moment is 5.5 times stronger then either alone. This brings the Protons magnetic moment up by a factor of five to around [$7.878 \cdot 10^{-26}$ J/T]. The Protons particle and Orbital motions fall in the same direction. The 2 magnetic fields generated from these two motions are coupled at the atomic level, and normally no external fields ever enter the atom strong enough to realign or break the coupling apart. They are seen as one complex motion magnetically in alignment, the result of two different spin frequencies, both in the MHz region. **Thus if we tilt the orbital field we also tilt the particle field.**

The Electron is seen spinning in the 28 GHz region. That's about 1000 times faster then the Proton. Its mass or weight is seen as very small compared to Protons [$m_p / m_e = 1836.152701$], yet the net electric charge is identical only reversed in value. The Electrons g-Factor is about 2, meaning that when its orbital and particle spins are coupled they add up to twice the magnetic moment [$-1856.9 \cdot 10^{-26}$ J/T]. These two spins are magnetically coupled as well and would both tilt together if suspended in a gas form. However in solid or liquid forms, electron orbits are fairly stationary forming bonds with other atoms.

These Two Relationships Are What May Give Us the Ability to Split the Fields

Magnetic Moment - Electron / Proton - Ratio: Ue / Up = 658.2106881

Mass - Proton / Electron - Ratio: 1836.152

The Electrons particle spin is seen in reverse of its orbital spin however, and this presents a very special case for a particle creating a magnetic field. The two magnetic fields are seen one winding around inside the other in reverse direction such that the particles momentum ends up reversed of its magnetic moment. This creates the particle of opposite effects when coupling to the Protons magnetic field. Electron particle spin is reversed of Proton particle spin if both are in the same magnetic field.

In a magnet these 4 spins all fall into alignment because over all they affect one another and loosely couple to one another orbitally yet tightly couple within particle orbital paths of each one. There are two kinds of magnetic alignment between these particles, and within NMR studies we see the differences. An orbital spin that receives excess energy through photon absorption, takes a spiral path and flips its motion completely over, reversing its N / S polarity. Normally when orbital spin flips, particle spin does also as the two are coupled together. To cause these two spins to decouple takes a very strong magnetic field far in excess of anything we could hope to accomplish in our garage. The energy is lowest when the magnetic moment is aligned with the external magnetic field.

The EM Field of Atoms

So here we see a picture where the Nucleus is hanging suspended inside the Electron shell attracted by charge and repelled in a cushion of magnetism, and coupled by spin through the magnetic field. The Nucleus a large mass with a small magnetic coupling, free to be flipped around by external magnetic fields which affect the momentum of the atom the greatest. As the electron orbitals are normally locked in place by chemical bonding or crystal structure, after magnetization they are pretty solid. Held in place by a much stronger magnetic force and far less actual mass.

As particles communicate through ESR and NMR they do so between "**like**" particles. Electrons may send photons that will not interact with Protons and also the reverse is true, yet each will exchange photon energy with their neighbors. Electron and Proton magnetic fields will tend to precess around the same vector, only flipping there alignment between two stable states. In both alignments 1 force of the 4 is always acting in opposition. An atom with a high nuclear spin will function very well at spreading NMR between atoms, as will a paramagnetic atom will with ESR.

Coherent Matter

Within matter at the core of all atoms, the nucleus, is setting the major mass of the atom. It is spinning, it also may have a magnetic field. In most normal matter the directions of spin is not aligned and will tend to cancel. Coherent matter is seen where the spinning mass of all atoms are moving in sync or in the same spin plane. There is another synchronization possible with a magnetic nucleus as well, where all magnetic dipoles are spinning in sync. This would result in a synchronous spinning magnetic field. In an object at rest, this coherent spin of mass may not be perceptible. It is only when we set this object in motion that we would observe anything unusual. If we can couple to this spinning weight and manipulate it, possibly we can cause device rotation, as well as bring Nuclear energies outwards. Coherent matter is where one or more of these atomic qualities are coupled across a great many atoms and they all begin to act as one with respect to at least one force. Magnetic, vibrational, electric, or mass rotation planar angle of spin.

Nuclear Mass Rotation

It should also be noted that it is possible to rotate almost the entire weight of a material like copper by magnetically spinning its Nucleus. If a great many of the Nucleus's of atoms can be set into a coherent spin plane of momentum, then tilting this momentum spin plane is the equivalent of tilting a fast spinning fan. If you hold an electric fan straight outwards, blowing away from you, then quickly tilt it downwards, you will realize the importance of this principle. It generates a rotational torque that tries to spin you around on your feet. This mechanism of tilting a spinning mass by rotating the spin plane applied at the Nucleus of atoms would impart about 95% of the mass of the material into a torque. This is very possibly why an induction motor is seen to turn in a rotating magnetic field. Copper setting in a rotating magnetic field will align its nuclear spin to the field, rotating the field will rotate the angular momentum of the mass of the copper atoms much like the fan in the above illustration. The spin momentum will sit at 90 degrees to the poles of the field. In the Roshin and Godin disc we see the horizontal magnets pulling the nucleus very quickly to one side, then releasing them to slowly return to vertical. During the fast tilt at 12 or more points along the cylinder this is causing an induction motor effect for a quarter of a rotation. If the spin momentum is strong enough the nucleus may continue to spin on around due to its own momentum. If the iron in the Searl disc is imprinted with a sine wave at the NMR frequency of copper, times the number of rollers, then here we might see one reason the device would both spin the momentum of nucleus into motion as well as reach a level of coherent matter. As we slide along the side of the Searl disc cylinder, we would see a spinning magnetic field imprinted over the vertical static field.

If we spin the copper nucleus magnetically, we get a physical torque. If we increase the frequency of magnetic spin so it hits the NMR frequency or a lower harmonic, which is the natural mass vibration rate, we may see a point where the nucleus begins to add energy to the momentum from its normal spiraling flipping motion. Coupling photon absorbed energy into physical spin.

The Neutron and the Strong Force

As we put the nucleus together, its weight does not equal the sum of its parts. The force existing as the "Strong Force" holding the nucleus together is removed from the atomic weight as a subtracted mass. Add up the weight of all the Protons and Neutrons, then subtract the gluon force converted to mass by $E=MC^2$. The weight is always less. The mass that is lost in the resultant atomic weight of the nucleus is the strong force or energy. The strong force is a sucking inwards of all particles inside the Nucleus. It draws them inwards.

Here we see a direct conversion of mass into binding energy. When the Strong force operates on the nucleus particles it reduces their mass by a finite amount.

This seems to be the first level where gravity is naturally altered within the atom. This may be where it can be controlled.

Classically three kinds of energy were identified that can come out of the nucleus, Alpha, Beta, and Gamma radiation. Either from parts exiting the nucleus as Helium and being expelled, or from orbital jumps to a lower energy level inside the nucleus. This is called nuclear energy and is far stronger then electron generated energy. Also the gluon energy is stronger then magnetism but only operates at a very short distance from the particle vortexes. This may give a clue as to how the particle vortexes work. Any atoms with more then 83 Protons are too big to remain stable and slowly fall apart over time. Their nucleus decays. Bismuth is the largest stable atom. How is it we can get two particles close enough together for the gluon force to grab them? Why does this process cause a weight loss?

Getting 2 Protons and an Electron together in a stable link to form a Nucleon, where the oppositely spinning vortexes overlap one another in the Neutron is amazing indeed. It requires reducing the force that holds them apart. Both are charged oppositely and so attract. Yet the magnetic fields are opposite of spin momentum in each one.

Orbital Structure of an Atom

Electron and Proton naturally attract from voltage [Fig 1]. When they sit in a magnetic field they both spin opposite directions. However as they come closer the spin of momentum coupling overcomes the magnetic field attraction. As we see in copper induction experiments, momentum is coupled through the magnetic field without being effected by the polarity of the magnetic field. Electron and Proton flip into a magnetic opposition [Fig 2], aligning their momentum vectors and a balance point is reached where voltage is pulling and magnetism is pushing. This can be achieved either vertically as pictured or laying broadside as in the Searl disc. We now have the electron cloud forming at some distance from the Proton shell. Indeed we see these two particle magnetic fields as very different, and this may be why matter does not simply disappear as we combine Proton and Electron. This pattern would seem to be of a scalar opposing nature magnetically at the orbital level. This may explain why the electron does not crash into the Nucleus, but why does it circle in an orbit? This answer may lie within the quantum physics model. As the Electron is only appearing at quantum intervals, and also spinning, the effect of this spinning pulsating force would be seen as wheel gears, turning against the ones the Proton is emitting. Talk about "Searl disc" similarity. These two spinning opposite directions with all the weight in the Proton, the Electron is hurled around the Nucleus at near lightspeeds. No more senseless talk about momentum type orbital paths in a frictionless space, this is all electromagnetic and powered from the spin of the Particle vortexes winking on and off as they cross this density.

Three Forces of Interaction Between Electron, Proton and Neutron
Spin – Magnetism - Electrostatic

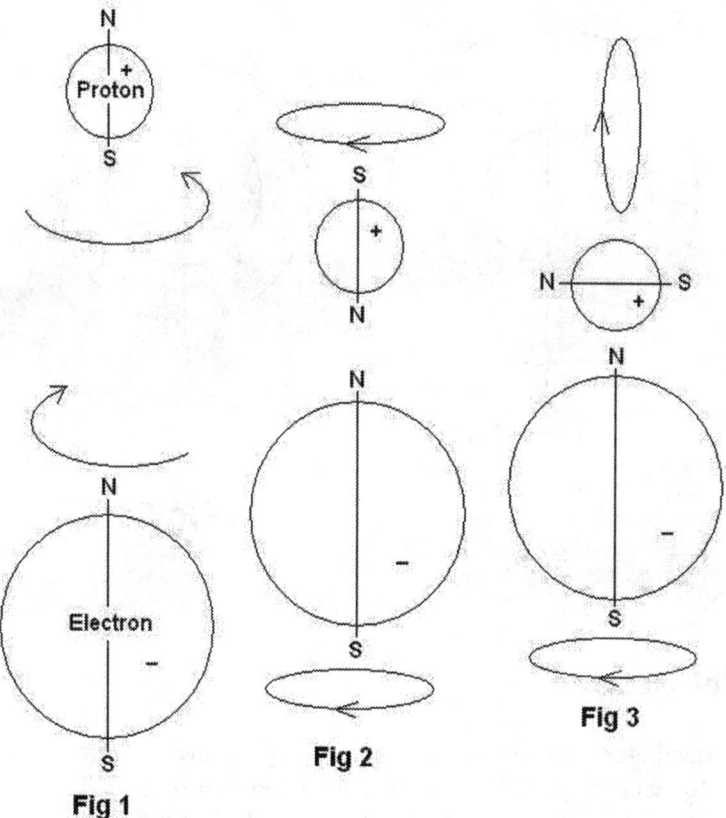

Fig 1

Fig 2

Fig 3

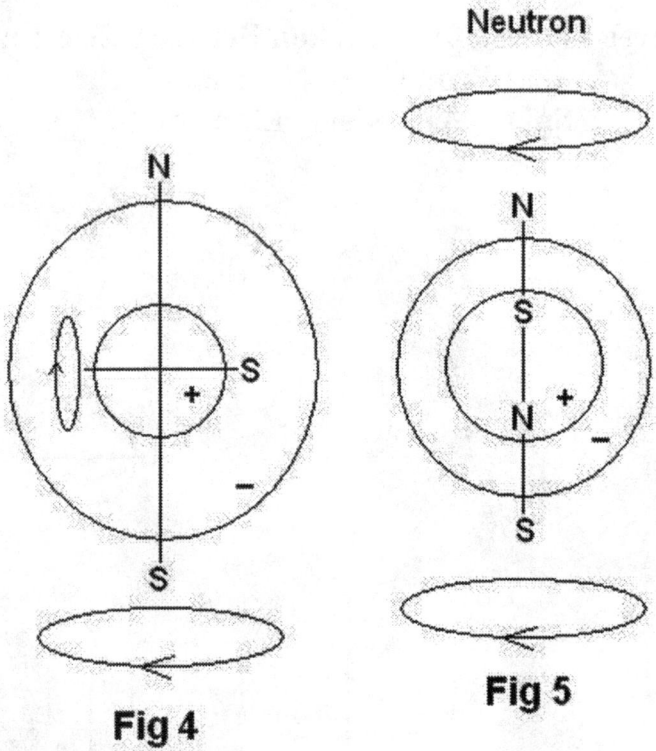

Fig 4

Fig 5

The Neutron

If quadrature alignment is reached [Fig 3] then Electron and Proton can merge [Fig 4] to become a Neutron [Fig 5]. The two forces of spin and magnetism setting at 90 degrees, the electric force pulls the particles together and overlapping. A beta energy particle is exchanged, and one quark flips. At this point their momentum swings back into perfect alignment, canceling voltage and almost all the magnetic forces from radiating [Fig 5]. [The Neutron still has a magnetic dipole moment of -1.913 Un. and there is a technology called Neutron scattering using its Larmor frequency as well, around 1 Gig]. When they are still setting at 90 degrees only the voltage vector is interacting [fig 4], the spin and magnetic vectors have not yet recoupled. It is between these two last steps where we may find the method of tapping the energy of Source.

As a Neutron they have become a scalar canceling device, falling into a momentum locked position with spins aligned. Interesting to note that only in this configuration do they now appear to have an attracting magnetic field, an attracting Voltage, and spin alignment. This unique alignment with one inside the other. This could be likened to a short circuit at the source generation point for the EM field. They are as close as they can get to all attractive forces. Both alter their size, the resultant Neutron size lies between Electron and Proton. A new force emerges in holding these together, and becomes the Strong Force coming now out of the Neutron just a little, sucking all the Nucleus together and altering the weight of the particles setting inside it. We now see a particle with a high mass and a weak magnetic field. The two tornadoes of Source "merging" as the strong force.

The Nucleon

If the Electron and Proton were to merge perfectly they would overlap exactly and completely cancel out one another, a total short circuit. This is very probably the instable state for a Neutron, a state of perfect alignment and may be what will cause it to fly apart. The presence of the extra Proton setting next to the Neutron in the Nucleon attracts the Electron outwards and pushes the inner Proton inwards [see fig 5]. In this state the Neutron is stable and can exist in all matter. If we were to increase the negative potential around this atom it would tend to push the Neutron closer towards this state. The higher Electron count would pull on the Neutrons inner Proton and it would push on the Neutrons internal Electron, making the Neutron less stable.

When the Neutron is removed from the Nucleus it will find itself setting in a more negatively charged field, if not entirely surrounded by the Electron shells of other atoms. To stay together it is noted that one Proton must be close enough to maintain a dominant Positive voltage with a dominant Proton spin. This forms the Nucleon.

If the Neutron is removed from the atom it will decay within 10.3 minutes back into a Proton and an Electron plus a Beta energy packet and a quark flip. Here we see a weight gain as well as an extra energy gain as they come apart. This is a surprise because we were taught that "energy in" must equal "energy out." A Neutron will dismantle itself outside the atom, giving off energy as it becomes heavier. How can a Neutron falling apart, both gain weight, and emit energy? So here we see a greater output then the sum of the parts with respect to Neutrons exiting the atom and exiting the Strong Force area.

The Neutron only produces the Strong Force when setting next to another Proton in a Nucleon. As a combined particle, the Neutron is only slightly larger then the Proton, and about half as large as an Electron.

Neutrons are not stable and can be manipulated, as the surrounding Protons keep them stable by providing spin coupling through the strong force as well as positive voltage potential and a weak magnetic force. This would tend to push the Neutrons inner Electron and Proton inwards closer to center of spin. The Electrons orbiting also effect them because their magnetic field is far greater then the Protons and it is opposing so also pushes inwards. This causes the Neutron to become smaller then the Electron alone and pushes it towards a higher density. This may actually be what alters gravity and the Neutron may be the particle that is the link to other densities. Here at the Nucleus all the known forces would seem to come together.

Momentum of spin - Electric - Magnetic - Scalar - Strong Force - Gravity

The Neutron can be viewed as a very close configuration of the same Electron and Proton particles we have dealt with in their Orbital positions. The close bond is one that is barely in balance within the atom, and comes apart if removed from the atom. It is very likely that we can affect this close relationship to extract energy from Source by developing a process of partially splitting the two while still inside the Nucleus. If we can toggle between Fig 4 and Fig 5 without destroying the Neutron.

Notes:

[An electron shell may form around Copper or Aluminum with its Proton magnetic field synchronized through all the atoms to become strong enough to repel the electrons magnetic field if induction is operating. A strong Proton generated magnetic field should act like the nucleus of an atom, if charged positively. The charge can be induced into the coppers electron shell but the magnetic field must come from the nucleus. Protons may orbit a negatively charged magnet as well.]

[Here we see that to produce the strongest possible Proton magnetic field, would require a material like Copper or Aluminum be charged positive. Then Protons aligned through NMR resonance into the same plane of spin. Hitting the system at 90 degrees with a strong electron generated magnetic field, we see the fields separating in the Neutrons, and tapping the strong force. This is the force that is shown to alter gravity right on the periodic table. This is also a force that is based on the two constants of spin found in the Electron and Proton which can never be depleted.]

Atomic Force Breakdown

Strong force - Overlaps all the Nucleus and particles inside it - [Neutron + Proton = Nucleon] [Effects mass or gravity lowering weight]

EM - Works mainly from the Protons outwards and overlaps Protons and Electrons, however within the Neutron there may be scalar canceling EM of the highest magnitude possible, where both forces are seen to create torsion or time at their closest distance.

Gravity - Seen as a constant force across all the particles, however at the nucleus within the strong force it is lower then the sum of all the parts outside the Nucleus.

[The Proton is seen as the Particle that seems to sit within both forces, interacting with complexity in both fields.]

[When a Neutron comes apart, breaking back into a Proton and Electron, it emits energy, as it leaves the Nucleus it gains weight]

Decoupling or Splitting the Electron and Proton Magnetic Fields

[The 90 degree tilt, reaching a new stable state between Proton and Electron fields, **Quadrature magnetization**.]

If we succeed at creating two magnetic fields setting at 90 degrees to one another where Protons align to precess around one, and Electrons align around the other, we could expect magnets to saturate along these two 90 degree angles as more atoms take on this alignment forming longer chains. We could also expect that the transmition of ESR and NMR energies would become polarized along such a 90 degree relationship as photons jump between atoms. Further the hint from the Searl disk that high voltages may accompany each field of the polarity following the particle generating the field. When the forces of EM no longer cancel within atoms and begin to appear at the edges of the material, we can expect a lot of energy to become present.

The Rain Maker Device

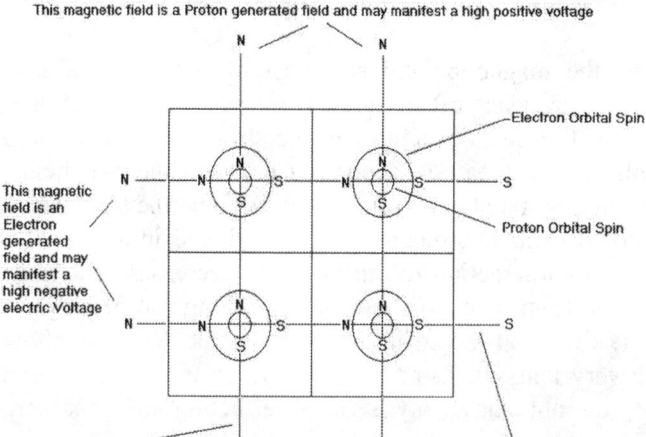

RF and microwave energy should radiate off the magnetic directions of the Electron or Proton orbitals seperating out into two 90 degree vectors. This energy hopping between adjacent atoms, tending to aid allignment.

Two Natural States of Alignment Between Electron and Proton Magnetism

State one: [The low energy state]
The magnetic moments are aligned - The angular momentum is opposing

State Two: [The high energy state]
The magnetic moment is opposing - The angular momentum is aligned

Either state sets **at least one force in opposition** to the other three.

Once the electron shell is bonded into a solid, the Proton field normally takes on one of these two states, precessing around it, unless a higher magnetic field becomes present at another angle moving through it.

Splitting the Fields

We see the magnetic field strongest off the ends of a dipole magnet and weakest off its sides. If we can place the **blotch wall** of each of the fields exactly over the strongest field of the other, we will get the greatest decoupling between the two fields spin planes. In this state the two fields will become the least interactive and should split. If a magnet becomes linked in a long chain of magnets, its interaction to others fields decreases. The magnetic interaction is strongest off the ends of a long run of magnets and near the center of the string it is very weak. You can show this with 2 very long chains of magnets, which if held together at the center, do not show any sign of affecting one another. Yet through each one is traveling a very strong flow and upon reaching the end of the magnet it jumps out to interact. A string of Neo disc magnets stacked to a foot and a half long, can be placed between two tables and will exhibit great strength at staying together. Upon placing it centered over another stack crossing at 90 degree angles, the two centers can be touched and moved apart and will neither attract nor repel. It is only where the magnet strings ends come together where we get strong interactions. Once quadrature magnetization is established over a certain distance we might expect it to be self sustaining.

The Rain Maker Device

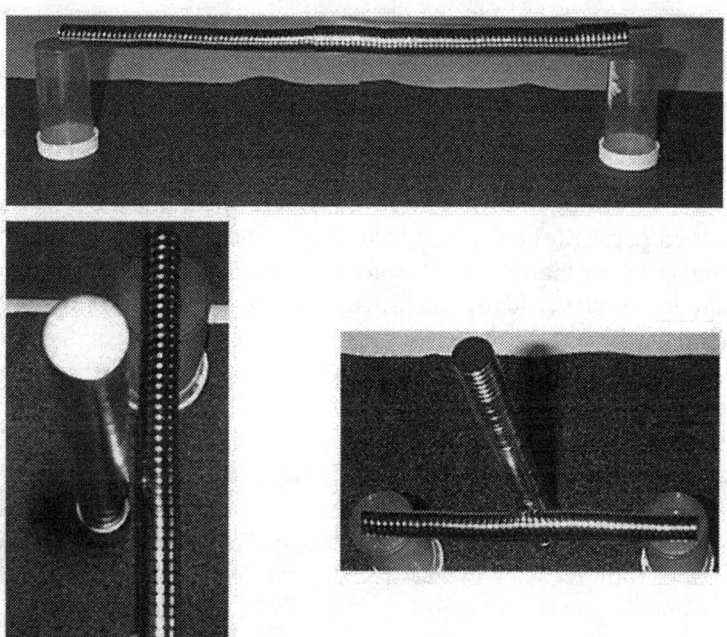

Long Strings of very powerful Neo magnets do not interact at the centers. Here magnetic fields can cross at 90 degrees to one another with almost no interaction.

In splitting we need to offer the string ends a good termination point. We want an Electron based magnetic field crossing a Proton based field at 90 degrees. Then we want to force the Protons and Electrons in our material to shift, chaining into the field that is operating at its like spin properties. All we have to separate them is the difference of weight, magnetic moment, and magnetic dipole precession frequency.

If we have achieved quadrature alignment at the EM layer then its force should be felt at the strong force layer as well hopefully affecting the Neutron.

Using a Magnetic Material to Split the Fields

One possible method [cobalt magnets]:

Since the Electrons 2 magnetic fields are usually fairly locked in during magnetization in a permanent magnet, if we position part of it horizontally and part vertically, then as we add momentum to the disc our Protons in the horizontal plane will want to align vertically. If we then coax them using another slightly weaker magnet at 90 degrees [vertical], they should tilt vertical and begin to precess around it, at some RPM where the momentum of the Proton begins to exceed its weaker magnetic holding force. This static state where both particles are at 90 degree magnetization is the one that may begin to cause relativistic alterations at the Neutron layer. This could be as simple as two magnets carefully aligned on both sides of a cylinder wall. At some critical velocity the field would split in the horizontally polarized magnet.

Another method of causing momentum in a magnetic disc:

If we simply place many stronger magnets at 90 degrees with moving intersecting fields such that we cause the Protons to all spin one direction along many points around the cylinder, we are effectively creating small rotating gyros. As we spin them through a 90 degree spin to the cylinders outer surface they will all impart a momentum to turn the disc. This is possible because nearly all the mass of an atom is in the nucleus, and it is the nucleus that will be spinning. You can show this principle by holding a fast turning bicycle wheel on its axis, while setting on a swivel chair. As you roll the wheel one direction in front of you, perpendicular to spin, one hand moves up as the other moves down, your chair will start to turn. While this method should be able to start a disc into motion it will not split the fields totally unless it spins them at 90 degrees to the primary magnetic field, yet it will create a small time interval where blotch walls pass at 90 degrees which may cause a smaller effect, but certainly not a complete saturation. This would appear however to be a **rotating magnetic field at the nuclear level**.

Using a Non Magnetic Material to Split the Fields

In a non magnetic solid material the electron orbitals are locked by chemical bonding and crystalline structures, however they are not magnetic. A material like Aluminum or Copper will need to be magnetized along its electron orbitals by passing a steady current through them one direction or laying them next to magnetized materials, if we are to couple them to the Neutrons inner Electron vortex. This will create a grouping of electrons along the correct spin planes simulating a magnetic atom. Both of these materials have a high nuclear spin so the Protons will be affected by both magnetic field and momentum. The key to this method will be to create the proper magnetization in the Copper horizontally, which will turn the electron orbitals into little magnets creating the 90 degree field. A second vertical magnetic field must be present slightly lower to capture the Protons as the higher momentum of the disc begins to tilt them vertical.

This method could be used by sheeting a vertically polarized magnet with Copper or Aluminum or even wrapping it with a coil as a torrid. It could also be done with only copper coils. Either way it will still have to be spun up to reach the threshold where Protons tilt vertical and Electrons stay horizontal. As this process is happening within the copper atoms, all can be started with DC currents. If it is desired to do this without spinning the disc, using only copper coils then some method of entraining the Protons with an NMR frequency may prove effective. The energy in the NMR would have to exceed the threshold of the difference of the two coils field strength. This would add a third coil pulsing the NMR frequency in the correct plane to tilt Protons towards the weaker field.

The Searl disc may be using many of these methods at different RPMs.

Kosol and Koeun Ouch, Vince Panella and David Lowrance

The [Avalanche] Effect
[Bringing the Micro Effect to the Macro World]

As we start this interaction in one small area of the device, it may be hard to envision how it can spread to envelope the entire disc, as well as effect surrounding matter. This is explained in the coupling between atoms setting next to one another as the two magnetic fields begin to increase in strength from the decoupling and from greater numbers of protons lining up in the new field. One of the methods of energy transfer between adjacent atoms is NMR, however this method is in fact frequency dependent. This means that as some of the Protons begin to tip to 90 degrees, their field in this direction becomes stronger. They will also radiate photons of the correct energy to effect nearby Protons tending to tilt them into alignment. **These photons will be ignored by the Electrons which have a frequency about 1000 times higher.** As new Protons get hit with NMR photons they will begin to flip into a spiral path towards reversing to 180 degrees. As they cross the new 90 degree magnetic field some will be trapped by it and begin to precess around it instead of continuing there flip. The Protons will all begin to seek the new alignment as time progresses, particularly if their NMR energy level is high overall. The Electrons will have a similar process tending to strengthen the horizontal magnetic field and holding them even tighter with the reinforced ESR energy oscillating between atoms in the plane at 90 degrees to the Proton energy. Also as the unpaired Neutron within Copper begins to split its magnetic field, the strong force will release energy, and the Electrons magnetic field will pop out aiding the polarization. As the process continues it may at some point become self sustaining, and spread beyond the 90 degree magnets of the device all through the entire system if configured correctly to allow this.

Two magnetic fields aligned at 90 degrees and building towards light speed unity. Particle velocity should speed up but never actually hit light speed, because the magnetic field is not perfectly dipole in nature. There is always some magnetic field off the sides of a dipole, however as frequency raises the tilt angles of NMR and ESR narrow, as the precession frequency raises, and far less energy will appear off the sides of the dipoles. This may push electrons into a high enough frequency to begin to radiate light. Thus the glow.

This describes a runaway condition we see in the devices studied. The better we succeed at splitting the B fields of Proton and Electron the faster the device may runaway. We can either attempt to slow the acceleration effect using very thin cylinders, or we can counter it with a method of reuniting the fields. Either way it must be accomplished within the device at the atomic layer before the device crosses the density threshold of 3rd density and vanishes.

Using Nuclear Spin to Decouple the Magnetic Flows

In studying NMR technology it becomes apparent there is another method of decoupling Electron and Proton fields, at least temporarily. When one hits the atom at 90 degrees with a 1/4 wave NMR pulse, the Proton field tilts to 90 degrees then slowly recovers, sometimes taking several seconds to stabilize back into alignment as it spins back up through a spiral path. With this splitting method the Proton field can be kept in a state of equator spin around the standing magnetic field. If a constant supply of pulses are fed to the material keeping Protons spinning then this alignment may be enough to alter the Neutrons. A standing magnetic field must flow through the material in the vertical plane, then pulses are added at 90 degrees in an attempt to keep Protons spinning on equatorial paths. If the pulses are hitting at properly timed intervals this method may be useful for thin materials such as cylinders or spinning cylinders. No DC holding field is possible for Proton alignment as it must be free to spin, so pulses must be kept constant, strong, and very short in duration, preferably four times the NMR frequency, or 1/4 wave pulses spaced exactly several wavelengths apart. Or a lower frequency wave form changing at the exact times to intersect these pulses edges. No stable quadrature magnetization could be expected to support itself in this method however it may be much easier to achieve if one uses good electronics designs. A formula for device cylinder spin, and pulse frequency could be set up to cover almost the entire surface of a disc with pulses before Protons can complete their realignment. This would establish a large enough area on the cylinder to generate high voltages between its outer and upper surfaces.

An AC holding field may be possible by setting up a rotating magnetic field at the NMR frequency around a thin material like a sphere or cylinder, with two iron core electromagnets above and below it. An iron core could even be placed through the center of the cylinder to transmit the DC magnetic field constantly along the Aluminum or Copper surface on the inner side. Are we starting to look like a Searl disk yet? The beauty of this design is that if we remove the pulses the effect should cease. If we decrease the intensity, then less atoms would tilt and a governor may be designed to regulate voltage output. Whether or not we achieve a total Nuclear resonance through the device would be hard to say, and experiment is in order. If the atoms stacked vertically on one another, do effect one another by magnetism and NMR energy this field may spread up and down the vertical sections of the material causing all to spin in sync at some point.

How Does This Effect Gravity?

Important to note here that it is a physics constant we may be effecting, that of Electron and Proton particle spin and relating to Planks constant. The Electron and Proton at their smallest points of spin in the Neutron may be the link to time and gravity, not so much the orbital motions. This is where we hit the wall of relativity as well as quantum physics. However since we gain control of their magnetically coupled orbital motions we have gained the ability to tilt them as well. This may be the link between the forces of EM and Gravity we have been looking for. As the Electrons momentum is less then 1/1000 the Protons this unlinking may seem of no consequence, however remember in a particle traveling at .999999... light speed this slight difference is magnified by a relativistic amount. Decoupling Electron and

The Rain Maker Device

Proton Spin momentum may be the strongest effect we could hope for in this sense. Decoupling them inside the Neutron may effect time itself.

As the Protons motions are freed from the Electrons motions from the decoupling of the magnetic fields now setting at 90 degrees to one another, both particles would increase in frequency due to at least one countering force being decoupled. Previously when they both sat in an attracting field, their magnetic moment is alike but the angular momentum of each one is opposite in direction. In the Neutron they sit with momentum aligned and magnetic fields in a powerful opposition. As the momentum is transferred across the magnetic field this would tend to lower angular momentum of both particles. As they decouple spin increases. If time is truly the increment of spin as Wilbert Smith has suggested then the overall time frame of the atom in this split configuration will speed up, approaching or exceeding the time flow rate of free space.

Time flow rate across an area of space is linked to the vector sum of all spin rates for all particle motions [Wilbert Smith]. And this is why the entire device as well as all surrounding mater within the corona become altered as device quadrature magnetic saturation is reached. The Searl disc shows us this phenomenon manifesting in a spherical area surrounding the device. Even dislodging the dirt in the ground if it is formed within it. At the corona appears a wall where time flow changes rather quickly, and this barrier may causes a torsion force that will rip matter apart.

When the time frame reaches an equal rate as free space we could expect the effected atoms to become weightless. This is a very small change in time flow rate and observed in the blue shift seen in the GPS satellite system. If there is enough energy in the interaction to push further then we could expect the object to become gravity repulsive and a push towards the next higher density. If the density threshold is crossed and the system then shut down, we may expect it to achieve the next higher stable state of matter, naturally landing it in the 4th density as the atoms fall back into normal alignment at the next higher frequency band. At this state it becomes totally invisible to 3rd density beings.

How much force must be applied to the Proton with its weaker magnetic field and higher mass to tilt it?

The requirement is only that we have enough momentum of spin to tilt the Protons magnetic alignment from the stronger field into the weaker field. These two fields must cross at 90 degrees inside the atoms. If using a non magnetic substance for splitting, the 90 degree field will affect the electrons motions as well, so it must be kept as much weaker as possible, otherwise the electrons may begin to realign around it as well. It is very important to find the threshold point and no more where the Protons tilt. As the chemical bonding of these materials do not hold the magnetic field within their Electron structure. As the EM field moves into quadrature magnetism it will affect the Neutrons and begin to split them as well if encouraged with a high negative charge, effecting time and gravity.

With the cobalt magnets however, both fields could even be the same. As the orbitals are magnetic within the structure of the materials there is little chance they will demagnetize one another at the same Tesla rating. However with both fields the same it may become impossible to return them to their original state. They may tend to toggle randomly causing a chaotic condition. The best interaction will be if the momentum is used to cause the threshold switching, as then the maximum possible numbers of Protons will be under device control.

Forces That Pull the Proton Vertical

Device spin, centrifugal force, mass, [aligned angular momentum]

Vertical magnetic weaker field. NMR in the correct plane of motion.

Force Window

The device spin [particle momentum I] must be calculated as to our [disc radius R] and our RPM [rotational velocity V] as an [energy value J]. It must be calculated to cross the threshold of [magnetic field coupling Bh] in the horizontal plane and the added [energy J] must be enough to couple to the [magnetic field Bv] in the vertical plane without crossing it. The parameters of [Disc Radius], [Disc Angular Velocity], and our two [magnetic fields Bv Bh] must be determined before we can expect a successful result.

These are the basic relationships that must be considered to arrive at a mathematical solution for initial startup.

Neutron Splitting

Getting the split EM field to penetrate to the Neutron layer requires more then just splitting at the Orbital layer of the atom. Neutrons come apart naturally when removed from the Nucleus. This includes a self tilting to decouple Electron and Proton enough to expel them back into magnetic repulsion states outside one another. One of the forces involved is charge. The Nucleus must have a high positive charge near the Neutron to keep it balanced. This is a force we can effect, and it is suggested that where a high - voltage builds up on a splitting material may be where Neutrons begin changing mass to energy.

Critical Threshold

It should be understood that this effect of new atomic alignment will manifest as a total **quantity** of the sum of all atoms attaining the new alignments, yet at some point these fields alone become strong enough to self radiate through the material of the device. **Thus a different method of control must be accessed to reverse the process or control it.** As well, the startup device must reach this **critical threshold**.

The **critical threshold** will be the point where the Protons new magnetic field becomes strong enough to maintain its new alignment without the aid of our external magnets, and it begins to spread outside the domain of the magnets starting the process. There will be a minimum number of atoms involved in maintaining the 90 degree field outside the magnets inducing the effect. These atoms will be linked by the 90 degree magnetic field the sum of all atoms with electrons in alignment and the NMR energy moving between the atoms.

It becomes apparent that each material we consider using must be calculated differently. This is because the nuclear spin of all atoms is different. Some atoms do not have a nuclear magnetic field we can even couple to externally. Others have a very strong one. However it is interesting to note that even the ones that do not radiate far may look like opposing magnets setting in one alignment.

The magnetic moment of the Proton field is readily available on the NMR charts as "magnetic moment" for most all the elements. It represents a Joules/Tesla or J/T value.

Nuclear magnetic moment of only about 2% of Iron is .15696, Copper is 2.87549, Nickel .96827 These numbers represent the magnetic field strength of the Protons magnetic force, they are different for each element and why NMR can be used to detect elements. The electron ESR is not so easy as finding it on a chart however. They must each be calculated based on the Proton field they sit within and the external magnetic field from outside the atom.

Control

The next most important issue after discovering the start up "critical threshold" is the control mechanism. With two magnetic fields manifest through our device, and two planes of ESR / NMR exchanging energy between atoms, the only solution is to reattach the magnetic fields one atom at a time and try to control the **quantity** of atoms in each alignment.

One method identified by Searl is to hit the device with a 28 to 29 MHz RF fields. This should tend to send Protons into an

NMR flipping motion of a higher energy state where as they rotate or flip they cross the electron 90 degree field becoming trapped in it once again for a time. The time this can last can be over one second as NMR has a time lag. While tilted back they will no longer add to device spin.

This frequency will have to be device dependent based on the material in effect and the standing magnetic field present. It could be something as simple as two resonant coils setting vertically and horizontally with a switch to electrically couple them. It would absorb NMR from one spin plane and then transmit it into the other, causing the Proton field to tilt as NMR interactive photons are radiated in the wrong plane dispersing Proton alignments. This would only work if NMR resonance is consistent across the device strong enough to couple to one of the coils. If it is not then an external transmitter must be used only hitting the band close enough to effect NMR flips. This transmitter must be located within the system such that it does not loose the time frame of the device. So control may become a balance of controlling the NMR energy between the two polarized planes within the device. Note this system is separate from the startup system, which can not be expected to control the process once started.

Materials

In the selection of materials it is noteworthy to identify a few very basic qualities of the atoms we are working with and attempt to correlate the parameters to accomplish a method of engineering devices, down to magnetic force splitting into the Neutrons.

Iron: The magnetic field in iron is mostly in its Electron layer. Of course the weight of iron is mostly setting in its Nuclear center which has a neutral magnetic field. When coupling a magnetic field through Iron almost none of the atoms weight or mass is affected in the least, [2% natural abundance]. As iron is magnetized, its chemical bonding connections are seen to roll with the external magnetic field setting up a new angle with other magnetic domains around it. The resistance to this force is linked only to the electron mass which is some 1800 times lower then its Nuclear mass, and its resistance to chemical bonding roll, and the temperature. A magnetic field generated from flowing current through Iron wire would be seen to have only an Electron generated field.

Copper: The magnetic field in copper is setting only in its Nucleus, attached to its weight or mass tightly, and able to roll if acted on by an outside magnetic field. Coppers electron shell is magnetically neutral. A magnetic field moving through copper is seen to couple to its "mass" and the Proton generated magnetic field will have to tilt the atoms weight as it changes angles. There will be a delay because of momentum during the tilting process as the weight will resist altering its spin angle. A magnetic field generated from flowing current through Copper wire would be seen to have both an Electron and a Proton generated field.

Aluminum: In Aluminum the Electron shell is also neutral, however its Nuclear magnetic moment is higher then Copper. Although the mass of Aluminum is lower then copper the magnetic coupling is higher.

Silver: I added Silver here because it has almost as low a Nuclear spin as Iron. With Silver wire one can generate almost completely Electron generated fields. Due to the fact that Silver is a better conductor then copper, these qualities make it perfect for the ideal splitting device if one needs magnetic coils creating magnetic fields with almost no mass interactions.

Cobalt: In Cobalt we see that both electron layer and Nucleus have a magnetic field inherent within them. An external magnetic field moving through cobalt will affect both. The electron shell is tightly bonded, so if tilting the Nucleus with a stronger magnetic field, it should decouple from the electron shell.

Effects: Effects that can be used to alter the free floating Nuclear Mass motions of materials are not limited to magnetism, or electric potential. Even though if you shake a block of copper atoms, it is the electromagnetic forces that keep the Nucleus centered and spinning, this force allows a flexibility to appear between the Proton and Electron shell. The Electron shell is extremely light by comparison to the Nucleus of an atom, but in solids the Electron shell is anchored into the structure bonds and its mass is far greater then one Nucleus. The Nuclear mass can be vibrated using many forms of energy.

Sound
Physical vibration
Moving magnetic field
A Vibrating voltage [Capacitance coupling]
Gravity

It should be noted that where the Electron Mass of a material comes into natural harmonic ratios with the Nuclear mass, a coupling of vibrational energy may be possible, if the material is suspended in a magnetic field and free to vibrate. [See section on Hamel Cones]

Quadrature Magnetization

If we wanted to produce a **quadrature magnetization** with the least amount of effort, it would seem the best material would be one known to have both a magnetized Electron shell, as well as a high nuclear spin to begin with. Of the three basic magnetic metals, Iron, Nickel, and Cobalt, Cobalt jumps out as the perfect candidate. We would expect that Cobalt would be the easiest material to create the effect in, as it naturally has both Proton and Electron magnetization. This is providing we can achieve a controlled separation of the two and align it to affect the Neutrons.

Moving towards metals that are non magnetic at the electron layer, we are now faced with causing them to become magnetized or placing them beside Electron magnetized materials like iron. The metals Copper and Aluminum jump out here as they are electric conductors and can be magnetized simply by passing current through them. If the current is passed vertically then a natural spin would manifest in the correct plane to align properly with our vertical magnetic field. The current would tend to tilt the electron fields exactly where we want them. We would still need the correct device spin to hold the Proton orbitals vertical, however now we have a device that may not need its 90 degree magnets at all. Simply wrap a magnetic cylinder with copper as a toroidal coil. Cylinder is magnetized N/S vertical, the same direction we flow current through the wires. The Proton tilts should manifest within the copper wire. This explains a great many devices at this point, combining magnets and copper wire in such a method. Now we may find some ability to define the correct parameters.

[For reference here are a few elements]

Iron -
Specific gravity 7.87 [mass/volume]
Isotope 57
Symbol Fe
Name Iron
Spin 1/2
Natural Abund. 2.11900
Magnetic Moment 0.15696
Gamma (x 10^7rad/Ts) 0.86806
Quadrup.MomentQ/fm^2 ---
Frequency 1.379 MHz
Reference Fe(CO)5

Iron is a natural material to use to place a magnetic field next to another material, however **it will not be very useful as to splitting the magnetic fields**. We see its natural abundance at 2%, this means we will get erratic results if any, as only 2% of its nuclear magnetic fields can be coupled to consistently with RF. There will be a very low amount of NMR jumping between atoms. Irons Nucleus would be impossible to align to a weaker field crossing at 90 degrees to its Electron field, and only device spin could do this. **Iron would be best thought of as a magnetic conductor able to distribute the magnetic field in a constant manner along a better substance for doing the splitting.**

The Rain Maker Device

Copper -
Specific gravity 8.95
Isotope 63
Symbol Cu
Name Copper
Spin 3/2
Natural Abund. 69.17000
% Receptivity (rel. to 13C) 0.06500
Magnetic Moment 2.87549
Gamma(x 10^7rad/Ts) 7.11179
Quadrup.MomentQ/fm^2 -22.00000
Frequency 11.290
Reference [Cu(CH3CN)4][ClO4]

Here we see copper is a better choice for splitting the fields. Its natural abundance falls into two different levels on the NMR chart. The most commonly found Isotope of copper is at 69% abundance [listed above]. Its next is found at 30% where we see a slightly higher NMR frequency. This means if we have a cylinder of copper built it may have one or probably both Isotopes found in varying degrees with two NMR frequencies present. Although copper is a good choice it is still a gamble and **two sets of numbers will have to be followed** assuming we will fall somewhere between them both. Copper is an excellent choice for coupling torsion from a magnetic field but **may not be the best choice** for splitting the magnetic fields consistently. Yet with a magnetic moment half that of Aluminum and a greater mass it may be far easier to accomplish with a wider **force window**. As copper is non magnetic at the electron layer, unless current is flowing through it, some form of current conduction will be necessary, or setting it very close to a magnetic material, to generate the horizontal field Bh [horizontal]. Current may have to constantly flow vertically through the material to magnetize it.

[The Searl disk is an example of the two Iron and Copper touching in a cylindrical method so as to make use of both]

Aluminum -
Specific gravity 2.7
Isotope 27
Symbol Al
Name Aluminum
Spin 5/2
Natural Abund. % 100.00000
Receptivity(rel. to 13C) 0.20700
Magnetic Moment 4.30869
Gamma(x 10^7rad/Ts) 6.97627
Quadrup.MomentQ/fm^2 14.66000
Frequency 11.095 MHz
Reference Al(NO3)3

Aluminum has a **natural abundance** of 100 %. This means that we will get very consistent results across all the atoms of Aluminum as found in nature. Its nuclear **magnetic moment** is very high 4.3 making it easy to couple to the Protons magnetic field yet a little harder to tilt it. In a one Tesla field we can expect NMR to be operating around 11 MHz consistently across all the atoms. Due to the high magnetic moment of Aluminum, we may run into a problem with RPM and the **force window** will be narrower. We will have to spin the disc around 2 times faster then a copper disc to break free of the horizontal field Bh. Another interesting note is that Copper and Aluminum lay very close to the same NMR frequency. Aluminum is not magnetic so it will be necessary to flow current through it to create field Bh if desired. It may also be necessary to support this field with 90 degree magnets because Aluminum is a poor conductor and creating a strong enough field to align Protons may be impossible without some external help.

Cobalt-
Specific gravity: 8.9
Isotope 59
Symbol Co
Name Cobalt
Spin 7/2

The Rain Maker Device

Natural Abund. % 100.00000
Receptivity(rel. to 13C) 0.27800
MagneticMoment 5.24700
Gamma(x 10^7rad/Ts) 6.33200
Quadrup.MomentQ/fm^2 42.00000
Frequency 10.103
Reference K3[Co(CN)6]

Amazingly cobalt seems to offer **all the needed parameters**. 100% abundance, high weight, high magnetic moment, plus it is paramagnetic, magnetized along the electron shell as well as at the nucleus. With this material we could be certain that both magnetic fields are present and can be coupled to for splitting the fields. Cobalt magnets run between 0.8 T to 1.1 T and should work for NMR Proton tilting easily. A cylinder rich in cobalt would need nothing more then to be magnetized horizontally Bh, and have opposing magnets at the correct Tesla rating Bv. There would be no need to have electric flow induced prior to tilting the Proton field. The entire disc could reach total saturation in both planes of quadrature magnetization. Two magnets of slightly different strength one as Bh and one as Bv a little weaker could be glued together or on opposite sides of a thin cylinder and spinning may be all that is needed to split the fields. Rows of these could be arranged around a cylinder of sphere, truly the simplest method I have found as of yet. Charging the Horizontal magnet with a negative HV would encourage a Neutron split.

Disc size

As to how disc size will effect the process lets consider a simple formula for angular momentum:

$L = MVR$
Particle Angular momentum = [Mass] times [Velocity] times [Radius]

This may represent the momentum imparted to a Proton or Electron setting within our disc.

We see that as we increase [Mass] [Velocity] or [Radius] we get a multiplying effect of the other two on the angular momentum. Increasing the disc radius can increase the effects of momentum on the outer layer of the disc by several magnitudes. We have no control over the mass of a Proton [$1.6726231 \cdot 10^{-27}$ kg]. We can however effect both the [Velocity] or RPM and the [Radius] of our disc. The suggestion in this formula is not to build our spinning disc too small, rather as large as practical. As the angular momentum of our disc will be coupled into the Protons angular momentum when they are sitting in alignment within a magnetic field, the higher we go the better. This is a coupling of angular momentum, and is not a coupling of magnetic spin frequency per say. Whatever momentum we are able to add will tend to spread our **force window** wider as Protons will be affected 1836 times more than Electrons. The difference must exceed the magnetic bond between Proton and Electron magnetic fields as well as the Electrons momentum alone.

Ref http://hyperphysics.phy-astr.gsu.edu/hbase/hframe.html
[Angular momentum of a particle]

Further if we consider the **moment of inertia** and apply it to a rigid spinning object [this is for reference and probably will not apply to just our particle]

$I = mr\char`\^2$
Moment of inertia [I] = mass times the radius squared
$L = I \times w$
Angular momentum [rigid object] = Moment of Inertia times Angular Velocity
From these two formulas we see a similar multiplying effect of [Velocity] [Radius squared]
Ref http://hyperphysics.phy-astr.gsu.edu/hbase/hframe.html
[Angular momentum] [Moment of Inertia]
It is noteworthy that doubling the radius of our disc should increase the angular momentum by a factor of around 4, reducing the RPM necessary by the same factor.

The Rain Maker Device

The larger the disc and the higher the RPM the wider the "force window" to reach critical threshold. Further to decrease the disc radius will require our 90 degree magnets to more precisely find the exact window where Electrons tilt and Protons do not tilt.

A Basic Smith Coil

Experiments with a coil similar to what Wilbert Smith described early on have shown the appearance of a 90 degree magnetic field. This field can be measured with a compass and produces one magnetic pole pointing outwards from the coils sides and the other pole appears inside the coil protruding from Both top and bottom ends. This seems to result from the coil canceling the normal Electron generated magnetic field.

The early description stated a coil wound such that each winding reverses the field from the ones next to it.

Winding this coil on an aluminum tube produces the 90 degree field, and when stimulated with frequencies between 100 KHz to 2 MHz produces a field that seems to interact with the nervous system of the human body. It can be sensed. It also is said to possess a conscious component when connected into the human brain.

Construction:

The coil was very simple and could easily be wound in a few minutes. I used about 12 feet of number 24 gauge insulated hookup wire forming only one layer. Starting at the center of the wire I looped the first loop around the tube, then as each next loop was placed on I twisted the wire on two sides of the coil such that one wire runs back and forth down each side. This produces a very uniform scalar field along the surface of the sides of the tube where the effect is strongest. When I got to around 1/4" of the end of the tube I joined the wires into a very tight twist about 2 feet long for feeding the device and then cut off the excess wire. This tight twist produces a good noise immunity at high frequencies, well into the MHz region without the need for shielded wire.

Since this coil is wound on either Bismuth, Aluminum, or Copper, it produces what I have theorized is a Proton magnetic field. As the windings bring the Electrons magnetic field into a state of contraction, or cancelling, the resultant tempic vectors simultaneously expand the Protons magnetic field in two dimensions. Since the Protons two magnetic spin components lie in parallel traveling coils to one another they do not operate the same as the Electrons, which both lie in opposition coiling directions. This opposition is the reason for the normal ballooning of the magnetic field. This coil seems to be contracting the Electron field to a point where the Protons magnetic field then expands out beyond it.

The properties of this new magnetic field appearing at 90 degrees to the normal field is the first solid evidence I have seen to indicate an alternate type of magnetic field does truly exist. When this coil is built around a larger copper tube and quickly hit with higher current pulses, that field can be mapped using a compass.

It will be interesting to see if this magnetic field is the cold energy we have been searching for.

The Magnetic Monopole

The magnetic monopole was observed by John Hutchison during some of his levitation experiments. Using the above coil designs if we wrap a Smith type winding over a spherical surface, one of two things should result. Either the magnetic field will loose its return flux path and shut down. Or one pole of the field will turn inwards from the Protons into the Neutrons space and stay in the lower gravity area of the Nucleus.

If there is another density lying directly inwards then we would end up with a magnet having one pole in this density and its other pole in a higher density.

Here it would become a Monopole. Since both magnetic vectors move the same direction in the Protons magnetic field, it should act like a one way flow. Either becoming an outflow or an inflow depending on which pole we turn inwards. By toggling the polarity we may be able to regulate the time spent in each state.

This sort of pulsing is reflected in almost all the devices studied to date. The Searl disc and the Sweet VTA are good examples that this model may be applied to.

With a monopole arrangement like this, it can be seen as a tempic pump, or a controlled flow of tempic vectors connecting between densities. The energy would be inexhaustible.

Comparing Other Devices

It may now become evident that what Hutchisen may have found was a method of entraining both Proton and Electron frequencies, and tilting them towards 90 degree positions, using the correct frequencies crossing such that sums and differences matched NMR and ESR rates in different quadrature planes. He did send objects into another **density**, which completely vanished, as well as levitate objects. The other possibility is to push both into higher rates of angular momentum by applying aiding spin forces to both simultaneously, this would seem all but impossible without first decoupling the two magnetic forces to some extent.

The Sweet VTA takes on a new outlook as well. Aligning the Protons in a weak magnetic field. It would take only a small force at the correct frequency to tilt the Protons into two canceling positions while entraining the electrons to flip the full 180 degrees. We end up extracting the electrons accelerating energy, while neutralizing the Protons energy. The reverse momentum of electron would be seen as the cold electric energy, in this model. As with normal electric flows, electrons are always flowing within the atomic orbits of atoms where greater opposing momentum constantly counters them. Also noteworthy that the actual splitting may be happening within the copper wire rather then within the magnet. The coils do not make any sense whatsoever as to turns to voltage ratios. The correct 4 coils set up with proper alignment may be all that is necessary, magnet conditioning may be merely setting up the NMR or ESR frequencies to split the field in the copper atoms of the coils. The quadrature polarization may be working its way even down the wires to the load. This would require the electron current to be moving parallel with the magnetic field rather then at 90 degrees to it. This energy would be delivering two magnetic fields rather then one.

In the **plasmas** we seem to see an energy gain, where electrons are completely freed from atomic orbits for a time in the ionized path through the air.

The Scalar Bismuth Core Coil

The Bismuth coil I have experimented with is about 3/4" hollow Aluminum tubes 1 1/4" long filled with Bismuth and wound with a Smith Coil which cancels the Electron induced voltages. However I have discovered the energy does not actually cancel but is reflected back into the Proton layer.

The scalar winding offers a mirror effect which allows a Proton - Proton interaction between two different inductive metals.

Induction Effect

We are familiar with the nature of induction with respect to Electron flow.

My version of induction, Electrons begin to flow through a medium like Coppers Electron shell. The flow is reflected back from the Proton layer as a countering force slowing the change of current. Inductance comes not from the iron core but from the magnetic Proton layer of the Copper itself. We see air coils offer inductance as well, and it is assumed that a coil in empty space would also offer inductance. Inductance is a function of the Copper itself operating between Electron and Proton magnetic forces which counter or attenuate one another lowering the energy in a circuit [back EMF].

Now in the Bismuth coil, we have set up a Proton interaction with another Proton layer in the Copper and a Copper Electron canceling layer. Any energy moving between the Bismuth Proton layer and the Copper Proton layer may pass through the Coppers Electron layer but will be canceled or reflected back in to the Coppers proton layer.

The Rain Maker Device

Bismuth is unique in that it offers a diamagnetic Proton magnetism that can be accessed outside or along the surface of its material base. It is this field in motion of its own nature that is acting in the scalar coil and being reflected back into the Coppers Proton layer that is powering the Vortex generator.

The scalar coil is the link between Bismuth and Copper Proton magnetism. The regenerative building effect is happening between the Bismuth and Copper wire that has its Electron field canceling. This is the true source of the interaction that is causing an energy envelope to pop out of the whole unit at approximately 3' diameter.

The surprise is that in the interaction of two inductive elements, pitting inductance against inductance does not create a degenerative effect but an over unity effect. The coil is self powered as long as the scalar winding is shorted. Opening the scalar coil halts the buildup of the field by allowing the Coppers Electron layer to form countering voltages that sink the energy.

Now with our coil we are hitting it with a diamagnetic field [Bismuth], and getting a counter field back at the Proton layer. So the Copper wire on the Bismuth core is a Proton - Proton magnetic interaction. This may be the effect we are seeing here.

Although a canceling coil tends to reduce Electron magnetism it seems to be doing something very different with Proton magnetism. It is actually forming an Electron energy mirror between the Proton layers of the Bismuth and the Copper where energy can flow between.

Placing the "South inwards" field along the coil is aligning all the Protons in the Copper and the Bismuth into the correct alignment for the Aether pumping diamagnetic effect which are jumping through the copper Electron shells.

The scalar Bismuth coil alone with winding shorted becomes the power source for the vortex generator of "Rain Maker 1" the first unit that I observed this effect in. It generates a field about 3' in diameter which builds on it's own over time and forms a "hot spot" in space which does not move when the generator is removed.

Kosol and Koeun Ouch, Vince Panella and David Lowrance

Hamel Cones

Hamel's concept of "mass into energy" would seem to apply to the Nucleus of atoms, where we see the Strong Force causing a similar effect of mass into energy. Here we see the atomic weight of the Nucleus is reduced by a mass and becomes the Strong Force related to one another by $E = MC^2$.

How is spin induced in Hamel cones?

The cones do not spin as we think of spin in a Searl disc, rather they vibrate, precess, or tilt, in little circles top and bottom. The vibration in the center cone moves in the same pattern found at the Nucleus of the atom in a magnetic precession, operating at several MHz in the atom. This is called the NMR [Nuclear Magnetic Resonance] frequency. Because the cones are interlocked, the top cones lower apex is moving in a small circle. The lower cones upper region is moving in a small circle with the same direction of spin but 180 degrees out of phase or tilt. The cone in the center is moving both top and bottom and couples to the Atoms Nucleus electrostaticly. The Nucleus of the atom is the heaviest part, and this coupling is a vibrational mass coupling of energy transmitted through the EM field between Protons and Electron shells. This is only possible because the cone is able to pass the Nuclear vibrations through its Electron mass structure, because it is also floating in a magnetic field and free to vibrate in this same motion. As the vibrations pass between atoms they cause a coherent NMR to link throughout a section of the Aluminum. Once the device spin or vibration becomes coherent, the heavier Nucleus of all Aluminum atoms will want to align with it in the same spin plane, just as with a spinning disc. The Protons and Neutrons will be pressed to flip vertical and there precession will be the same as the center cone. Along the magnets at the rim of the cone the Protons will be held in horizontal spin by the magnetic field and unable to flip upwards. At some point away from the magnets, as the magnetic field drops away, the

Protons and Neutrons will be found tilting upwards, overpowering the magnetic field. This is the point or ring of vibrational threshold for step one. The NMR frequency will determine where this latitude is located along the cones mass and the physical vibration will be found to be a subharmonic of the NMR frequency along this one latitude. This process is identified as step one, and is the source of the cones motion or spin. It is the device motor.

In my current model, the two step process for tuning a Hamel cone would be these:

1 - Achieve the highest possible cone vibration resonance frequency through good design choices. [get the cones to vibrate]

2 - Split the magnetic fields by finding the resonant subharmonics that land the NMR frequency closer to the horizontal magnets.

How does the Hamel cone split the fields?

The NMR frequency of Aluminum changes with the strength of the magnetic field it is setting within. Therefore as we move away from the horizontal magnet ring along the cones larger circular edge towards its tip, the NMR frequency drops with the magnetic field strength. In step one we cause one ring or latitude of the cone to enter a coupling with the Nucleus of the atoms and the cone starts to vibrate at some subharmonic frequency of the NMR frequency. There should be found that by increasing the compression of the field the cones are setting in, more then one rate of precession may be achieved, and the frequency of vibration should increase, moving the latitude of NMR closer into the magnetic field. If we can get the NMR ring to move into the correct area of the magnetic field, then across some small region it should cause the Protons to tip up where the horizontal Electron generated magnetic field is at the correct intensity to cause the Electron and Proton fields to split inside the Aluminum atoms all the way down to the Neutrons. This is step two, quadrature magnetization.

At the place along the cone where the horizontal magnetic field ratio is correct, if the Protons have attained a strong vibrational coupling along this same latitude, the field should split tilting the Protons magnetic field vertical, breaking away from the Electrons horizontal field. Working its way into the Neutrons, as the field splits within the strong force area, energy flow begins outwards, now through a vibrationally coupled system, all the way to the Proton and Electron layer and the split begins to propagate throughout the cones. Negative electric force charges the sides of the cones surface and somewhere around the center point we could expect a positive charge to manifest. Control is with a vertically polarized magnetic field, said to gate the flow. However this magnet would tend to upset the sensitive balance between vertical and horizontal magnetic fields, in such a thin cone material, and rejoin the fields by overpowering them both in the thin Aluminum.

The force window can be seen as getting the first NMR resonance to manifest on a cone latitude as close to the correct magnetic field area as near the magnets as possible. The vertical tilting force on the Protons, and the NMR coupling frequency latitude of cone mass to NMR frequency, must cross at the correct horizontal magnet strength to form a splitting area. If the two never come close enough then a cone that simply vibrates may never cross the threshold of splitting the fields and releasing Neutron energy. Getting the cones to vibrate is only the first step, this creates the vertical momentum force for Proton alignment. They must resonate an NMR subharmonic, with a mass physical vibration, crossing the magnetic field at the proper intensity to create the split of Proton and Electron magnetic fields along one latitude, before it can spread. It would seem that strength and placement of the base and upper magnets may help support the vertical field, and more attention should be directed here as well. Being able to regulate the gating magnets strength may allow a sliding of the NMR ring latitude slightly up or down the cone to coincide with the proper distance from the horizontal magnets to begin the splitting process.

The Rain Maker Device

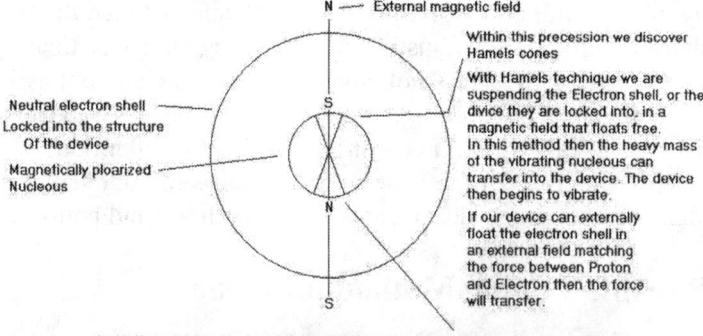

Aluminum atom has no magnetic field at the electron layer yet has a high spin at the nucleous Causing a high magnetic moment. In a 1 T field the nucleous will allign with the external field.

N —— External magnetic field

Neutral electron shell
Locked into the structure
Of the device

Magnetically ploarized
Nucleous

Within this precession we discover Hamels cones

With Hamels technique we are suspending the Electron shell, or the divice they are locked into, in a magnetic field that floats free. In this method then the heavy mass of the vibrating nucleous can transfer into the device. The device then begins to vibrate.

If our device can externally float the electron shell in an external field matching the force between Proton and Electron then the force will transfer.

The full mass of the nucleous will precess around the field

If we match the NMR frequency to the device mass vibrational frequency, as it floats in the magnetic field, by a harmonic function then energy should couple.

Startup can be tunned by measuring the free floating mass vibrational frequency of the cone and tunning it until it hits a harmonic of the measured NMR frequency.

The vibrational energy will first couple to the device and then to other atoms. All the Protons will begin to precess in sync at resonance with one another. Startup can be aided with a 10Mhz coil horizontally polarized at the exact NMR frequency. Control can be with a vertically polarized 10 Mhz coil to break up the resonance in the horizontal plane of motion.

If a lower tesla rating is used on the magnets then the new NMR frequency must be adapted to which may be much lower.

Since the atoms of Aluminum have no magnetic force at the Electron layer then probably very weak magnets can be used, however they all need to be of a very close rating, as any variance will effect the frequency pushing it out of device resonance.

If the electron mass of the free floating cone can be vibrationally coupled to the vibrational rate of the nucleus at a **subharmonic** then nuclear Resonance should envelope the cones upper and lower areas where circles are widest. If we could build a cone such that its electron mass was equal to the mass of one nucleus this would vibrate up rather quickly. This small cone would have only around 3,813 atoms total and its total electron mass would be equal to the mass of one Nucleus. Having identical weight it would easily vibrate at the exact rate of the Nucleus and begin to couple the Nucleus vibrations between atoms. Growing larger requires a sub harmonically coupled vibrational rate. Lowering the magnetic field will help because it will lower the nucleus precession rate. This will not affect the coupling mass but it will allow for vibrating a heavier object more quickly at a lower harmonic.

These parameters are within our means to measure and to easily alter as well as to control. The Hamel cone is tuned through a laborious process of adjusting the opposing magnets floating distance, altering the stress holding the mass of the cone, thus its vibrational rate. However if the cone has not been, by chance, engineered to fall within a certain range, then no vibrations will surface. It would seem the cone in the center is the one that will emulate the atom most closely, as it circles both top and bottom.

A Possible Tuning Method for Cones

The NMR frequency can be determined mathematically for Aluminum for the magnets used and the distance they are at from the Aluminum, dropping off at an inverse distance cubed rate. Two devices can be fabricated to measure both vibrational mass rate, and NMR rate.

Mass Vibration Sensor

A section of the cone is polished to shiny. A laser light pen is connected to the frame and hits the cone reflecting off onto a calibrated piece of white paper, with a grid pattern and a photo sensor behind it with a small hole. Cone vibration is seen as a dot size increase, or a line width. If the paper is moved further away this will be amplified. A simple circuit can connect to a frequency counter that should read the frequency of vibration.

NMR Sensor

A coil is wound around the frame, outside the cone of about 30 meters of wire horizontally level. The length of this wire could be near the calculated NMR frequency wavelength. This could be wound on a separate form that may be removed. A second coil is placed inside the cone with vertical polarization as an NMR receiver. It must be near the center of the cone and have a shield around only its outer edge, not its side edges. It can be fitted with a tuning capacitor and set to resonate at the NMR frequency as well. A strong signal generator can now be used to scan the cones resonance. When the NMR frequency is stumbled on a simple RF field strength meter should show the cross coupling between the two coils as a sharp resonant peak, or a group of them appearing along the cone. This can be achieved with a diode and a voltmeter much like Hutchison's hand held unit. The positioning of this coil must be very accurately vertical and the connecting wires must be very short and shielded running to the meter that must have a shielded box. Or a radio receiver could be tuned using a CW receiver, and following the signal generators frequency.

As the cone is frequency scanned in this manner the vibrational rates of both "device resonance" and "nuclear resonance" should show on one of our sensors. Finding the rates and then altering parameters to tune them to cross at one frequency harmonic is the method for tuning in step one. A fully tuned magnetically floating cone may physically vibrate when the correct NMR frequency hits it, indicating that the cones electron mass has linked to the NMR vibrational rate. It should be determined how many frequencies this can be made to happen at, and how high we can run this frequency up.

Magnet Selection for Hamel Cones

All magnets are a combination of Proton and Electron magnetic fields, which have opposite spin momentum in the same field. The horizontal magnets running around the cones should favor Electron spin and avoid Proton spin. Proton magnetic spin carries with it a high mass momentum, Electron magnetic spin does not.

The best magnets to use for the horizontal field would be magnets with little or no Nuclear spin. Magnets with a high Nuclear spin like Cobalt will tend to resist the Proton tilt in the Aluminum slowing the splitting process. Ceramic ferrous [ferrite] magnets would probably be the best choice. Ideally the Protons in the Aluminum will be held by the Electrons magnetic field in the magnets we select and easily tilt away from it when the spin force becomes strong enough. Gating magnets aligned along the top and bottom of the device may be Cobalt or even Neo. However this may just as well inhibit shutdown and control.

An alternate method is suggested for building the cone magnets using Iron or Ceramic ferrite rings. An iron or ferrite ring can be very precisely magnetized by spinning it past very strong magnets at high RPM. A very uniform magnetic field will result. Spinning the cones ring past a neo magnet South pointing inwards will produce a uniform field in the iron with North pointing outwards. The reverse can be done with the stator ring. The setup for this would require a spinning test jig able to spin the rings up either before they are attached to the cones, or spinning the cones afterwards. The normal method of coil pulse magnetization produces the wrong shaped field for a Hamel cone, resulting in a North South vertical alignment like the Searl disc uses. Testing can be checked by slowly turning the ring and watching field strength from a stationary point. It should be very constant along the entire ring.

Once they are magnetized, stronger magnets must be kept away from them because iron can be affected quickly by Neo magnets touching them. Alternates would be to do this with a machined magnetic material or a molded ceramic and magnetized after assembly. This would ensure a uniform field. Piecing separate magnets along the ring will produce a random intensity magnetic field and NMR may not find a complete circle around the cones. If separate magnets are used then each one should be tested with a compass against the earths field to ensure a same distance reach is present where the compass is seen to move half way between the magnets force and the earths force.

Adjustments

Adjustment of the magnetic compression screw should cross the two vibrations at several points, and the higher the better. Just because a cone vibrates, does not mean it will not vibrate at a higher, almost undetectable rate, if compressed further. The vibration sensor would be a valuable item to detect these higher rates and to peak them. Pushing the physical vibration rate higher

The Rain Maker Device

up towards the 16 KHz limit of physical vibration may help. Nuclear vibrational coupling will be increased the closer the frequencies come to one another. The Aluminum NMR frequency may be setting at around 1 to 5 MHz some inches off the horizontal magnets. A chart can be created to predict where the NMR rings would form and mapped onto the cones as a frequency. The cone must vibrate at a sub harmonic of the NMR frequency landing as close to the magnets as possible for splitting the fields.

An Experimental Device for "Splitting" in Aluminum:

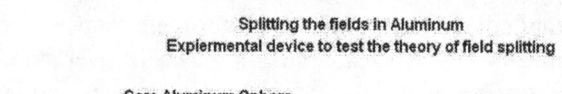

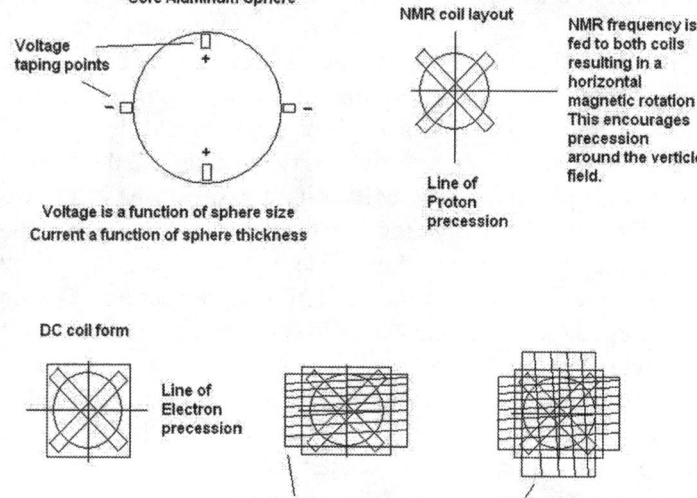

Control is through NMR coil phasing. If coils are fed 180 degrees out of phase by reversing the feed wires of only one, then NMR mag rotation shifts into the vertical plane rejoining atoms.

Concept of Operation

4 coils are provided for individual control of Proton and Electron motions. Device is designed to find the correct parameters by allowing a large number of variables to be applied. The DC coils are outside the NMR coils so they will not interfere with the high frequencies hitting the sphere. No iron is to be used in the device, because the individual magnetic fields will be skewed or bent as they pass through iron. All the magnetic fields must converge in tact inside the Aluminum sphere. The outer DC coils fields will pass through the NMR coils because they are not Alternating fields, and will receive no inductance as they pass through them. However the DC coils will attenuate some RF from the NMR coils.

It is noted that silver wire would be better then copper wire because it has a much lower Nuclear spin and thus coils overlaying one another would offer much less induction interference to one another.

The Electron bias coil is provided to keep a strong Magnetic field in the horizontal plane reaching all the way to the Neutrons. The Proton trap bias coil is a slightly weaker field designed to catch the Protons as they are tilted vertical delaying them from returning for splitting. The NMR coils are set up to offer two planes of motion induction for the Nucleus, one will split the fields and the other will join them. The difference is achieved by reversing one coil polarity, rotating it 180 degrees in phase. Also other phases can be experimented with for various tilting angles if phase driving circuits are included.

Measurements are available with a compass, and a voltmeter. The core sphere should develop a voltage as with the Searl disc when the fields split, + at the center and - on the outer ring. Voltage taps are provided for drawing currents and large copper conductors should be used. The compass may be kept near the top of the vertical North pole, and slightly to one side. As NMR resonance is hit, this pole may be seen to spread tilting the compass slightly downwards no longer pointing to the exact center. This is because the lines of flux have found new paths through the Aligned Protons coherently through the material. If

The Rain Maker Device

Aluminum wire is used to tap the voltage we may expect the field splitting to migrate up it and actually provide the cold electric effects. It is thought that by using copper the dissimilarity should stop the splitting from leaving the device.

The ultimate device would add the ability to spin the sphere, in the case that this may be the only way to get the Neutrons to separate, since the inner Protons magnetic field seems to be totally hidden under the Electrons. The Neutrons magnetic moment is [-]. It is my assumption that since the Sweet VTA does not need a spinning device, this experiment may be enough. I am expecting that since the orbital Protons are operating within the strong force area, tilting them will be enough to split the Neutron forces.

If quadrature magnetization can be accomplished in Aluminum, I would expect a constant voltage electric force appearing, the sum of many electron volts aligned at atomic levels, and probably a function of the magnetic chain length or the diameter of the Aluminums solid parts. Thus sphere size may become important to voltage output, and thickness may be important to current capability.

Device 2 - Rolling the Proton Field at an NMR Frequency

There is one other method to split the magnetic fields that may prove interesting, requiring slightly more energy input. That of rolling the Protons field through a circle positioned 90 degrees to the steady Electron magnetic field. If it can be rolled at an NMR frequency then Protons and Nucleus of atoms will be spinning in one field and the other field will cross it at 90 degrees. In this configuration the entire Nuclear mass of the splitting material is spinning although the device is stationary at the electron layer and does not appear to move. With this method iron can be used in the electro magnets in the stationary field, so can be much closer and stronger. The NMR coils will be situated at 90 degrees to one another and fed 90 degrees out of phase setting up a rotating magnetic field at the NMR frequency or a sub harmonic, or even a short pulsed waveform, also setting just inside the stationary magnetic field.

Proton dipoles will be held in a constant state of circular spin, in a spin plane perpendicular to the Electron field. An Aluminum disc or cylinder could be used for this experiment.

The **Nuclear Magnetic Resonant Battery** is a similar device concept.

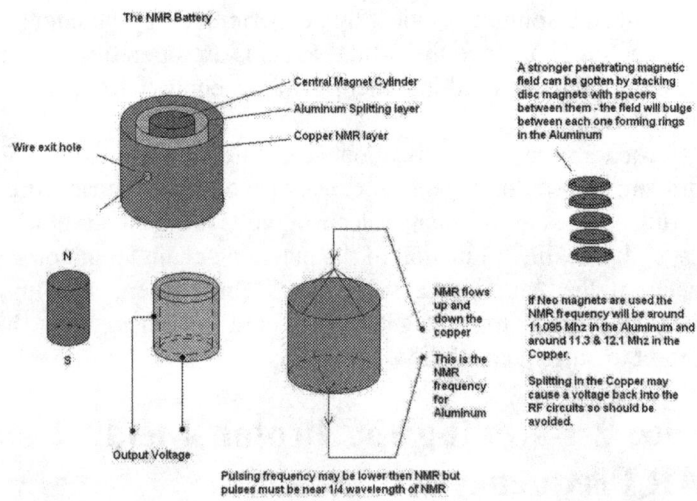

In the Nuclear Magnetic Resonant Battery or NMR Battery pulses are fed through a Copper tube or cylinder vertically. This creates a magnetic field pulsing around the inside of the tube at 90 degrees and causes the Protons to move into a magnetic spin, still precessing around the vertical magnetic field, but at 90 degrees to it. Because the Copper tube completely wraps the Aluminum tube all the Protons are affected at once and the device does not have to physically spin. When the Aluminum reaches Nuclear resonance at the full 90 degrees the fields split into the Neutrons. Noncancelling EM begins to appear along the surfaces of the

The Rain Maker Device

Aluminum top and bottom [one voltage polarity] and sides [other voltage polarity]. The device is totally controlled electronically by the pulse amplitude and pulse length. If the pulses stop then the Protons slowly spiral back up into alignment with the vertical magnetic field and it should shut down.

Magnets Dropped Through an Aluminum Tube:

The picture of me dropping a stack of Neo magnets through an Aluminum tube, is one of David Hamel's demonstrations. The magnets seem to float down the inside of the tube because the entire weight of the Aluminum Nucleus must rotate as the magnets move through it. Since the mass is also spinning it resists having to turn.

One further experiment should be added to this to point out exactly where this interaction is happening. Drop a short Aluminum or Copper tube, a couple inches long, down a long string of Neo magnets. You will discover that it only falls slower near the ends of the magnet, and falls freely down its center.

This does not mean that there is no magnetic field along the center of the magnet string. It only means that the angle of the flux lines are not changing.

Looking closer at the magnet floating down the tube, we see directly below the magnet an area in the Aluminum where random Nucleus angles are being aligned with the magnetic field. This area is a great resistance as all Proton spin is aligned first. Next we see the flux begin to roll as the lines cross a 90 degree point with the walls of the tube very near the magnets bottom end. This is a coherent roll of all the Protons in a ring around the tube together. Next we see the center of the magnet passing and there is no turning of the flux angle so little resistance is encountered as all the Protons are aligned and not turning. As the top end approaches we see another coherent magnetic roll in the Protons before they are released back to random flipping motions. This adds one more drag at the other end of the magnet. This is why only the ends of the magnet are causing the resistance to motion. Another note worthy factor is that as the magnet approaches the Aluminum atoms their magnetic spin frequency is increased. As it move on by the frequency drops back down.

An important concept is seen here, in that if we have a material like Copper or Aluminum in a high speed rotation setting in a magnetic field, as long as the field is constant in shape, and the rotation is perfectly in alignment with the field, there should be no resistance after all the Protons are aligned. Only the spin up will cause a drag. This would require a very accurate alignment of magnetic field with device spin, but results in a coherent Proton magnetic field throughout the material offering no drag.

Searl Disc

The Searl disc is seen as 4 layers of materials. Center is Nb, and said to magically create electrons. Next layer is Teflon and said to gate electron flow outwards. Next is Iron with a vertical magnetic polarization as well as an AC signature at some RF frequency. Last the outermost layer is copper. Around this perfectly formed disc are the rollers which are basically the same only with their vertical field reversed so rollers are attracted to cylinder.

When in operation there is a negative voltage charge developing on the outer ring. Power can be taped off by small pickup coils placed above and below the rollers. as they pass through mid air they induce voltages into the Copper coils evenly from above and below.

Searl Disc Magnetization Process

Looking at the way the Iron Cylinder is magnetized we see an RF riding on top of a Static DC magnetic field. Since the RF programming winding is narrow and placed at the center of the cylinder its flux lines reach out crossing the DC field at nearly 90 degrees only along the upper and lower edges of the Cylinder. An RF wave pattern is imprinted into the Electron magnetic domains when the Main field is placed into the Iron cylinder. Looking at the microscopic level what we end up seeing is a quadrature alignment of magnetic domains in the Iron in these two areas. The edges are peppered with atoms in each alignment. Some align around the DC flux field and some around the RF flux field. As the RF field is reversing [AC] its flux is in motion moving out then in and intersects in quadrature only at two areas near the cylinder edges. These are the only place where they may remain stable after magnetic coils are shutdown because everywhere else they are not decoupled enough not to effect one another. After magnetization these are locked in and act as little 90 degree magnets pointing straight outwards along the edges of the cylinders.

Operation

As the rollers move by these they act like atom sized spokes, interlaced with both quadrature polarities, the Roller grabs the Cylinder and must spin to roll along it. As motion speed increases the Copper layers of rollers are moving through the Cylinders magnetic field and begin to generate the familiar magnetic repulsion no matter the polarity of the magnetic fields they encounter. The Copper layer is generating a cushion of repulsive magnetism that the Rollers float on. Where 90 degree magnets are encountered these swing the Proton fields to 90 degrees in the copper where splitting must be happening at a level that can convert mass into energy in the Neutrons. When Coppers Protons encounter the magnetic fields generated from the Iron they spin couple, and take on a repelling magnetic polarity, so the Protons start to fall into a magnetic pattern reflecting the Disc programming in the Iron only repelling in the magnetic field.

We see the same scalar forces inside the atom in the orbital process.

As to why the rollers do not fly off at higher speeds, this requires some thought. The attracting magnetic fields holding them at low speeds are not strong enough to hold the rollers onto the disc at high speeds. At some point the centrifugal force will exceed the magnets holding strength. There are two or three other explanations as to why the rollers would stay on at high speeds. Electrostatic attraction is one, if the rollers become negatively charged and the cylinder becomes positively charged. Another is the Strong force found in the Nucleus of atoms. If this force is expanding outwards to encompass the entire disc then at 137 times the magnetic field strength it would have much more ability to hold the rollers. Also the Proton tilts may be generating an inwards angular momentum along the surface between the cylinder and the rollers. As these 90 degree tilts are turning the mass of the atoms [the Nucleus] then at higher speeds they would provide more inwards torsion. Of course another idea is that as the device lowers gravity then momentum drops along with it and the rollers actually weigh less so are easier to hold on. With all Mass in the device spinning in a horizontal plane of motion there would be no momentum resistance to horizontal movements.

Source Energy

Lastly it is speculated that when RPM reaches a high enough spin it will become impossible to tilt Protons at all with the magnetic fields crossing at 90 degrees. It is at this point where the only field that will be allowed to turn is the electron with its far lower

The Rain Maker Device

mass. The only Electron that is not device coupled, is the one setting inside the Neutron at the center of the atom, having only a very small part of its magnetic field protruding. If the pulsing magnets can reach this level, or if the negative potential becomes strong enough, it may be this tilt that taps the strong force and begins to release Source energy.

Device 3 - Simplicity

This device is the simplest that I can conceive to test the enclosed theories.

An Aluminum Cylinder mounted to a high speed electric motor with an insulated shaft.

A ring of high intensity magnetic coils are positioned around the ring as close as possible at 90 degree magnetic polarity.

A High Voltage power supply.

The goals are this:

1 - Align Protons and Neutrons vertically with spin momentum.
2 - Place a High Voltage Negative DC charge on the cylinder.
3 - Hit the Neutrons with a 90 degree high level magnetic pulse, causing it to partially split.

These should be under variable control:

RPM
DC voltage
Magnet coil strength, pulse width

    The Hamel cone shows us that it may not be necessary to even pulse the 90 degree coil, if the magnetic field strength is correct.
    The Searl disc shows us it may also be necessary to add a ring of magnets or a magnetized layer under the Aluminum,

however this may only be necessary to replace the motor, as a high enough RPM will put the Protons where we want them.

Our target is the Neutron which has about the same magnetic moment as the Proton. The Neutron however has a negative magnetic moment which is a result of the shrunken Electron shell setting around it. This shell would have a very low mass and should tilt much easier then the Proton setting inside it so this is our target.

The NMR frequency of the Neutron is just up into the microwave range, setting up a microwave transmitter is not an easy idea, however we do not need to resonate the Neutron only split its component fields. The Spinning disc should keep the Protons which are heavier locked in the vertical position, allowing the Neutrons Electron shell to tilt. Since in Copper or Aluminum the Nucleus has a non cancelling spin, there should be at least one Neutron with its internal electron field exposed. If we can snag this one and tilt it then we could expect to see a higher energy effect, then in messing only with the Orbital shells.

If we choose to add a vertical magnetic field to the Aluminum cylinder, it is important to get this field very well positioned. The magnetic fields N/S poles must be true to the motor shaft and not tilted or skewed in the least. A spinning tilted Magnetic field will generate all kinds of spin resistance due to its alternating intensities. A magnetic layer added on our cylinder must have a constant strength to any one stationary point. The reason this is so important is that we do not want Proton magnetism altering its spin angle as the disc turns.

As we experiment with RPM, DC voltages, and Magnetic pulses, we are searching for a magnetic pulse that does not add inductive drag to the Aluminum. As soon as an inductive drag is appearing this means we have tilted the Proton field and are tilting Nuclear mass. The goal is to stabilize the Nuclear mass of the Protons in the vertical position, and find the correct magnetic pulse that will reach the Neutrons negative magnetic moment, without tilting the Protons. The Neutron will respond to far higher frequency pulses then the Proton also.

David Lowrance
Public Domain Paper 03-15-06

Notes and NMR References

Note 1
In AG work it is common to use values relevant to a 1 Tesla magnetic field, as many Neo magnets land between 1 and 1.5 Tesla. It is noteworthy to mention however that the precession rates [NMR and ESR frequencies] given in this paper are relative to 1 Tesla and these rates will be altered as the field is lowered or raised.

Note 2
It is also well know now that the particles Proton and Electron are not really particles, but energy in a wave looping function. However as they still possess the same measurable effects of magnetic moment and angular momentum, I have chosen to use the word particle to describe them.

Note 3
A mental image presented of the Roshin and Godin Russian Searl disc duplication may be an aid to understanding references in this paper, as well as a basic understanding of how NMR and ESR function.

http://www.cis.rit.edu/htbooks/mri/inside.htm
http://www.rexresearch.com/roschin/roschin.htm

Other References

http://hyperphysics.phy-astr.gsu.edu/hbase/hframe.html
http://hyperphysics.phy-astr.gsu.edu/hbase/magnetic/magpot.html#c1
http://hyperphysics.phy-astr.gsu.edu/hbase/magnetic/magmom.html#c2

http://hyperphysics.phy-astr.gsu.edu/hbase/nuclear/nmr.html#c1

Definitions

Spin Plane - As an electron is seen in an orbit around the nucleus, any one complete circle lies within a single plane, or plane of motion. As the magnetic vector lies at 90 degrees to this it is important to have some way to reference the angle. This is similar to an electric coil, whose winding all lie approximately within one plane, and the magnetic field generated lies at 90 degrees to this plane. While the Electron eventually winds its way around the entire electron shell, its spin plane shifts, however if the atom is magnetic more of these spin planes align to form a greater field external to the atom along one spin plane. The electric spin plane is seen 90 degrees to the magnetic field and tends to contain more electron motion of one direction of spin.

EM - Electro magnetic force is 1/137 as strong as the strong force. EM is the force operating at the Electron layer, and between Protons and Electrons.

Strong Force - The force found at the nucleus of atoms where vortexes or particles are very close to one another. The Neutron is seen as an Electron vortex overlapping a Proton vortex such that spin is coupled and EM is canceled. In this tight configuration the strong force is dominant. It drops off just outside the nucleus and does not affect the Electron cloud. Strong force can only hold 83 Protons in a stable configuration. Bismuth is the largest stable atom. Within the strong force is a weight reduction where the total mass is lighter then the sum of its parts.

Nuclear - With reference to the unit at the center of an atom consisting of Protons and Neutrons which are tightly bound by the strong force. The major weight of the atom exists within this nucleus. The Electron shell is seen to exist outside the influence of the strong force, and operates only by electro magnetic force.

The Rain Maker Device

Proton - The Proton is seen generating the major magnetic force found in the nucleus of the atom, thus it is the coupling link between the atoms major mass and magnetism for atoms that do have a nuclear magnetic moment. Tilting the external magnetic field that is coupled to a Proton tilts its spinning mass rotation offering a resistance of momentum. The Proton is seen to spin the same direction of its orbital motion and as such offers a higher energy output then Electrons when magnetism and momentum are combined. Particle that spins CCW in a magnetic field, bound in the strong force at the nucleus of atoms.

Nucleon - A paired Proton and Neutron found in the Nucleus of an atom.

Neutron - The combined Electron Proton vortexes that cancel their Electro Magnetic fields, and are bound in the nucleus by the strong force.

Electron - While described as a particle, the electron is really a quantum wave energy moving between densities. It is only in our density at very specific intervals of time or quantums where its force is felt like the spokes on a wheel as it spins. The importance is that we have identified its main measurable manifestations. That of magnetism, electricity, and radiation frequency as related to "angular momentum" and "magnetic moment." The electron spins opposite its magnetic moment. Electrons create a magnetic field some 600 times stronger then Protons with a mass over 1800 times lower. Particle that spins CW in a magnetic field, freely roaming the electron cloud and is not bound by the strong force.

Quadrature - Forces that are linked together in a 90 degree relationship such that energy between angles of forces are linked. The electron set into motion creates a quadrature magnetic field. The Electrons spin on the quantum level partially creates the time flow rate [Wilbert Smith].

Quadrature magnetization - The theoretical state within an atom where Protons precess around one magnetic field, and

Electrons precess around another magnetic field, such that both magnetic fields sit at 90 degrees to one another. Both fields sit in the **blotch wall** of the other. This may be a necessary transitional configuration to the creation of Neutrons, where both are seen to end up overlapping.

Density - A stable state of matter, one set of parameters where the atom is stable to interact with other matter at the same state. Our physical world is one such possible state, where atomic particles have found an equilibrium. The sages through the ages have envisioned many more such worlds and named them. They indicate we are now on the 3rd leveled density of mater. Claimed Alien contact as well has suggested this model is accurate and moving into higher densities changes the relationships of momentum, gravity, and time flow rate.

Source - The sought after place where free energy is moving into this density and powering the atoms and all manifestations of energy found here. The regenerative source, or the root location of tapping that source of energy. I believe that Source will be located within matter, and is found at the crossing point of atomic particles entering and leaving this density.

Dipole - The quality of magnetic fields as they are created from electron motion to manifest two directions of force along its B vector. One labeled North and one labeled South. The earth has a magnetic South pole setting at its North Geomagnetic pole. The North pole of a compass, will attract to a South magnetic pole of a free magnet.

Blotch wall - The center point between the dipole ends of a magnet where almost no magnetic field exists with enough magnetic polarity to affect anything. Holding one magnet inside the blotch wall of another allows for the least interaction between them. Within atomic orbits that are more perfectly aligned spherically, this effect may lead to another stable state of atoms, once supporting magnetic fields are established.

The Rain Maker Device

Compass - The simplest most convenient way to detect a magnetic field, although often overlooked today as a tool of science. Caution is in order because if held too close to a Neo magnet may be completely reverse polarized if forced into a reverse field alignment. If using a compass it should be periodically checked against the earths field and realigned as needed using a neo magnet.

Reference Constants and Formulas

Constants
Planck constant: $h\ 6.6260755 \cdot 10^{-34}$ J·s
Proton mass energy equivalent: $1.503\ 277\ 43 \times 10^{-10}$ J 938.272 029 MeV
Electron mass energy equivalent: 8.187×10^{-14} J 0.510998 MeV [Me C^2]
Proton mass: 1.672621×10^{-27} kg
Electron Mass: $9.109\ 38 \times 10^{-31}$ kg 5.485×10^{-4} u
Mass ratio Electron / Proton: $5.446\ 170 \times 10^{-4}$
Mass ratio Proton / Electron: 1836.152
Proton g-Factor: 5.585
Proton magnetic moment: 1.410606×10^{-26} J T^{-1}
Proton gyromagnetic ratio: 2.675×10^{8} s^{-1} T^{-1} over 2pi = 42.5774813 MHz/ T^{-1}
Electron magnetic moment: -928.476×10^{-26} J T^{-1}
Electron volt-joule relationship 1.602×10^{-19} J 5.609×10^{35} eV = (1 Kg)C^2
Joule-electron volt relationship: 6.241×10^{18} eV = 1J
Proton Electron ratio: Ue / Up = 658.2106881
Electron magnetic moment: $-9.2847701 \cdot 10^{-24}$ J/T
Proton magnetic moment: $1.41060761 \cdot 10^{-26}$ J/T

NMR Formulas

Energy of a photon E = h V [h = 6.626x10^-34 J [Planks constant]] [V = frequency]

Gyromagnetic ratio $V = Y B$ [$Y = V / B$] [V = frequency [Hz]] [B = magnetic field strength [Tesla]] [Y = gyromagnetic ratio]

Transition energy $E = h Y B$ [$h = 6.626 \times 10^{-34}$] [$Y$ = gyromagnetic ratio] [B = magnetic field strength] [V is between 15 and 800 MHz typically in medical apparatus]
Ionization potential for an organic molecule is 6×10^{-19} J
In NMR the energy in a photon is a function of its frequency times planks constant. For the photon to flip a nucleus magnetic field over this must equal the **Transition energy**. There is only one B field strength that can match any particular photon. Atoms setting all in the same magnetic field will easily exchange photons.

Force Vector Model of the Atom
The Lastest Comment From The Technician
[David Lowrance]

A combination of intuition and science, reflecting the merging of two opposing concepts creating infinity and perpetual motion in the paradox we know as reality. In this presentation "We" begin to merge the spiritual and physical forms of reality at a deeper level. From the outer conscious fabric of awareness to the inner physical reality. I use the word "We" as I know that I am one of many beginning to awaken to the merging of the spiritual and physical sciences. "Illusion" is a word found to describe inadequate understanding of the innocence and purity of the universe.

View From the Conscious Fabric

In the beginning God was without experience. We are all a part of God experiencing itself.

The truth of God is found in the Tao. God is the source of all opposing forces, and God is found in all sides of all conflict.

The illusion of a polarized God is in error. God is in fact the [Source] of all things good and evil.

As mankind resolves itself into factions of like mind and emotion, it finds itself naturally in opposing groups of ideology on the fabric of perception.

"Light speed simply is". All knowing and all existence all at once, in the absolving of all conflict. To experience time flow, consciousness must slow light speed to drag it into a state of sequence and variable rates. To accomplish this consciousness creates the first force, the first force is division of itself into two equal and opposite forces attracting one another but cannot become merged. The first force creates perpetual motion that can never be satisfied or neutralized without transcendence, and this is the basis of the universe below the God realms. Motion that can not be stopped. This force is rooted in the Control fabric. There is nothing inherently within matter to explain perpetual motion, or infinite spin, or unstoppable spin that we perceive in atomic particles as reality. This is not willingly observed by those rooted in the physical sciences. Consciousness had no other option but to divide into two attractive forces of divided intention. When opposition became present then light speed could be slowed down to create separation. The tempic vector is therefore the first force acting against itself to slow light speed and establish perceptible Time flow rate.

Within the control fabric, "consciousness" splits itself into two opposing tempic vectors and turns them against one another, from the original intention.

This was the beginning of the yin and the yang and the infinite dance.

"After the conflict in the perception and control fabrics absolved into clear intention of two divided goals, God force moved down into the layers of space and decided to create forces standing in conflict of itself. Thus began the physical manifestation and the source of all the dances of motion began.

Consciousness found something called "life" in the unresolved conflict of opposite intentions. Polarized conflict is the source of all existence in a tempic field below God awareness. Thus "time flow rate" is rooted in the conscious fabric."

To do this consciousness must act on the conscious fabric with intention to create tempic force and slow infinite awareness so that it can experience itself in finitely small quantities of opposition. This is depth, consciousness slowing reality to sequentially look at each part of itself in finite depth. All possible combinations of forces are a part of Source. Consciousness creates force in the form of a circle so that all places it channels itself to can perceive the same rate of spin and reference events with a common sequence, however how can consciousness escape itself, so it must always return to the starting point. The circle powered by conflict of intention, constantly reversing, is the only infinite form of perceptible reality and the God force finds itself innocently ecstatic in the experience. The dog chasing its own tail. The pure innocence of God, in the beginning, and the beginning of force began in the indecision of direction of two intentions set in opposition and consciousness following both paths in a circle of increasing radius. The first spiral is born in the conscious fabric, and the expansion of the universe outwards, subdividing into progressively more conflicted intentions. Thus the longitudinal wave went out within the conscious fabric and sustains all.

Proton and Electron are separate and can not be joined but are held in perpetual motion by the fabric of Consciousness.

The first forces appear as circles of tempic vectors turning against one another and held in opposition by consciousness alone.

"The true God is not blind to suffering but the true source of everything found in the perception and control fabrics of reality. God is the source of conflict at the purest form because the opposite forces are in fact a part of Gods commitment to divide itself. This is the purest form of consciousness present. God is the source of conflict of itself in all aspects, and this is inevitable on all levels of existence. Inside us, a reflection of God, is the force of both powers, to be separate, and to be connected. This is the beginning of the "perception and control" fabrics and explains the depth of the rest of the forces. This choice was not taken lightly by God but found necessary for consciousness to express all the possible expressions of realties so that perception could experience all of them."

Conscious Fabric

To move towards Source [God] we must absolve conflict at each level.

Two opposing intentions slow the time flow rate and expand time into sequentially perceptible chunks of opposition. "Outflow" Life.

Two like intentions speed the time flow rate and create attraction, resolving conflict and drawing towards Source oneness. "Inflow" Awakening.

Thus in the conscious fabric there is no death only life and awakening.

Life is rooted in "paradox". The primal intention lying at unity is one and only one "intention" the intention to create division and conflict in the fabric of consciousness and thus all the structures of the fabrics as they expand outwards and slow the time frames of awareness. Thus on the conscious fabric time flows more slowly as we move outwards from Source unity intention. The single point focus is the absolute slowest moving time frame, the peripheral awareness, grasping the greater picture all at once, is the absolute fastest moving time frame and embraces all opposing intentions. Thus slowing the mind increases the knowing. "Illusion" is resolved by absolving conflicting thoughts. As the control fabric is a variable balance of inflow and outflow, time flow rate is not constant across all consciousness. The more we embrace conflict of intention the larger the conscious fabric becomes, the more we are involved in the expansion and the slower we progress through time. Pleasure is found in the reversal at both turns. The human collective has embraced conscious conflict on no less then 7 levels and thus honors the prime directive of life to explore every possible conflict of intention on every possible level of existence. Awakening is the undoing of the prime directive to end conflict at each level and move back to Source. In the cycle, life finds infinite satisfaction.

Consciousness sits in quadrature to the physical.

The atom is God's perfect infinite conflict set in the physical world to operate exactly opposite of how it appears to operate.

Within this "perception and control loop" we find a paradox powering it, perpetual motion induced into physical matter. The ecstatic pleasure found in the reversal between conflict and unity. The game of life which begins in the Perception and Control fabric and is the prerequisite to the physical plane. It is based in the "emotional" fabric of the God force and creates the astral plane of awareness, the playground of life manifesting physical reality. Thus the conscious realm is divided into the two levels, one of mind and one of astral emotion. The reversal of consciousness following both paths creates the loop of the tempic vector around the God force alternately seeking both at different intervals of times to form a loop on itself between the perception and control fabrics experiencing outflow and inflow from source. Thus we find at the roots of matter all particles are in flux between the inflow and the outflow and none are present consistently. The path they take is the conscious path and lies in quadrature to the physical plane.

Physical Fabric Paradox

As we examine a circle in physical space we observe the first contradiction to the control fabric. For a tempic vector to travel in a circle it must turn more slowly on its inner side and thus time appears to move more slowly towards the center of the circle then along it's outer edge. This is the only way that a vector of motion can move in a circle and thus the first reversed perception is born, that center is dead and time is frozen at the center. This is how matter appears yet stands in complete conflict of how things work on the levels of the Control fabric. From this observation we must conclude that the operation of matter is reversed from the true operation of the universe and that Electrons physically setting outside the Nucleus are in fact a closer unity and thus time flow rate is higher as we move outwards in the physical plane. Thus a higher density is one with increased physical radius. Embracing a larger physical awareness brings unity in the control fabric and thus truly speeds the time flow rate. However perception of this is reversed in that as we actually move to a higher time flow rate we find a unity of intention and the physical universe appears to shrink from the perception fabric layer. This demonstrates that the Control fabric is the true reality and the physical is the secondary reality in opposition. As we observe inflow and outflow sensation this reversal is also true. While standing on Bell Rock in Sedona Arizona one is expanded in consciousness sensing physical out flow and yet simultaneously drawn inwards to God consciousness. While setting in the inflow of Cathedral rock one is draw physically inwards only to discover every intimate detail of the causal events of each and every conflict they have ever experienced within separateness from God consciousness and personal Karma.

As consciousness moves outwards to embrace all physical matter the time flow rate increases, and as consciousness moves inwards to the single point in physical space it becomes lost in the infinite flow of time as time flow rate decreases and even seems to halt.

The physical fabric lies in quadrature to the conscious fabric, however perception is reversed as to its operation. As Source divides itself outwards in the conscious fabric, it spreads itself out over all the physical realm moving inwards towards a physically stationary center frozen in time. This creates the illusion of entropy. As physical space expands, consciousness slows the time frame of a single point awareness. This is the nature of the tempic vector appearing oppositely in both fabrics.

The meditators have shown us the conscious fabric and the scientists have shown us the physical fabric and both sit forever in paradox and contradiction until an expanded awareness is available to explain and relate them both.

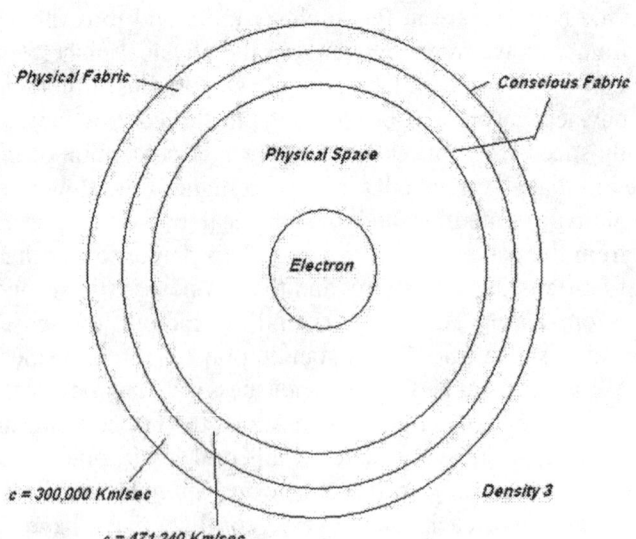

The Rain Maker Device

From the above we realize that gravity is a force pulling us physically inwards towards a slower time flow rate and was placed by intention to separate or divide consciousness into many forms and possibilities all separate in infinite variety, and thus we observe a universe of infinite variety located in small pieces of God awareness with a small single point awareness. To unite diverse galactic cultures is to move awareness back outwards and begins the reunification of God force itself inwards. This begins on the social level and culminates in the merging with God consciousness for all, but the formula is always the same, recognize conflict at all levels and resolve the diverse conflict of intentions.

Space was the first creation of consciousness to allow for the separation of individual conscious forms in opposition of intention.

Consciousness traversing space is the beginning of reunification and drawing back to Source. Thus the draw to move outwards physically and begin to explore the space that has separated all consciousness into separate pockets of [Source] dividing itself. As consciousness moves outwards in the physical it is moving inwards in the conscious fabric of perception and control. When all separate pieces of consciousness once again reaches the boundaries of the universe then all will become God and the universe will be done with its experience of itself.

For now, consciousness finds attraction in the perception of opposite control forces manifest as beauty, at perceiving itself in opposition of intentions in infinite variety. This is the prime directive of Source and nothing else, to cycle between the outflow and the inflow.

A machine to traverse density therefore would act to physically expand matter at its deepest level, the Neutron, turning the parallel physical forces into what appears to be physical opposition, which is in actuality unification of intention causing the true size of the universe to shrink along the conscious fabric. Only through physical expansion of the smallest level of awareness can we shrink the universe to a size that can be quickly traversed.

Kosol and Koeun Ouch, Vince Panella and David Lowrance

From Kosol:

The Guardians said that ZPE or ZPM now has been made by David Lowrance using their guidance.

The Rain Maker is that ZPE and ZPM system.

Once the consciousness unit is created by the Rain Maker, it is active constantly. All one has to do is put a crystal over it, but the crystal must be wrapped with regular coils. The energy from the consciousness unit is transferred into the crystal spherical medium and the coils wrap around the crystal medium converts the spinning magnetic and static charge into unlimited useful electricity.

Here, take a look at Dave's Rain Maker device using Kosol and Guardian concept.

Ok, all Dave needs now is a coil on the crystal, because the consciousness unit he created caused the crystal on that device to have static charge and all Dave needed is a coil on the crystal to siphon the energy from the static charge on the crystal to the coil, also you can use aluminum plate as well plus the coils. There you have it, you will get infinite energy from this ZPE, ZPM, and Rain Maker device.

Kosol,

Welcome home!

I will have to give this a try, it sounds very good.

I have been deep in the next part of defining the conscious fabric, and wrote the last piece about "The vector model of the Atom" on the Alternate energy site.

The Rain Maker Device

I know it may seem pretty deep but at last it feels like I have connected some dots as to how time flow really works and what God is really up to.

LOL!

I will give this some more thoughts but I am thinking a spiral coil or flat pancake coil to convert the inflow outflow directly to electricity. But I must try all suggestions.

I also have yet to try a different layer of iron pipe instead of the TV ferrite rings to see if that simpler form will work to make the device easier to construct.

Thanks Kosol,
Dave L

Creation of the Physical Fabric

I have sensed that if I keep going with this conscious awakening, I may experience once again awakening on another level, and the whole physical world may vanish in an instant as the playground I have created, just as the Grey's have vanished when I awakened as to the Astral planes true operation. Truth will often appear as insanity to those rooted in the physical plane. The following is an exploration into the very fabric of creation, and thus may hold the key to it's undoing on the personal level. Only the brave soul should move this direction when it is ready to be shaken loose from the physical fabric. In my present awareness I have perceived this presentation to be the focal point between Science and Spirituality, and shall allow the crossing of mathematics between the two. I also view this as a natural progression of the work of Wilbert Smith as offered over 50 years prior, and the beginning of this process of the merging of science with spiritual reality and entrance into the greater brotherhood of all the alien races.

Creation

[One of the strangest and hard to comprehend passages that Wilbert Smith left us. "The boys upstairs tell us that, Light simply is."
    This is the greatest clue as to the true operation of the conscious fabric.]
    No matter how complex a phenomena may appear, it must conform to the prime structure of the universe.
    The two primal forces emerge from the original intention [prime directive] of Source consciousness to divide itself outwards in conflict with itself.

The Rain Maker Device

The inflow and the outflow are the conflicted intentions to move consciousness either away from or towards Source and alter it's perception of time flow rate. All other forces are a derivation of these 3, [inflow, outflow, and tempic]. Consciousness thus establishes itself in the inflow outflow loop as pieces of the God consciousness divided from Source. Consciousness moves at light speed and in this sense "Light simply is". Light speed is not fixed and slows as consciousness moves away from Source. The Perception fabric sits in quadrature to the Control fabric.

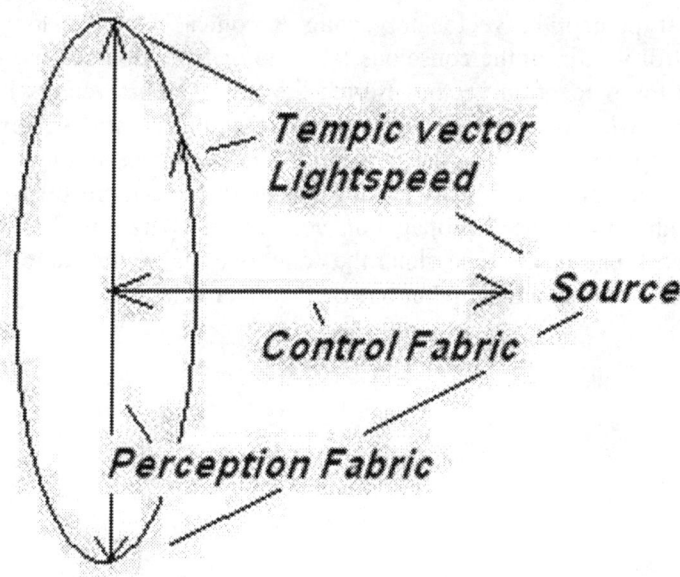

All force is derived from the two original vectors of "intention" slowing light speed at distance from God awareness, so opposition can be perceived in variable levels of detail or density, and creates the Control fabric [decision] moving away or towards Source to alter the time flow rate. The "outflow" is the intention to move away from Source along the Conscious fabric, and the "inflow" is the intention to return to Source. Outflow moves away from Source along the conscious fabric so as to create a pocket of slower perception of time flow [slower light speed] in a given density. Consciousness now spins "perception" around itself at that lower light speed to create the [appearance] of motion and an inverted or reversed tempic vector. This is also perception seeking to alter its path in a circle to perceive all sides at light speed.

Perception is moving at light speed, and perception slows as we move along the outflow away from source. Perception moving in a circle slows the perceived time flow rate even slower through repetition and allows much greater time to observe the interaction. This appears as the quantum packet or the time that matter appears in this density to be present, and takes the shape of a spiral perception vector traversing a conical section along a control vector on the conscious fabric to or from Source. We see it is the perception vector offering the three tempic vectors in 3 space, what has not been apparent is that the control vector is altering light speed as the perception vector moves through our density. Thus our density would appear to have more then one possible light speed propagation velocity in operation, different on each end of the tube along the cone, one on the outside of the conical section and a higher one on the inner side.

The Rain Maker Device

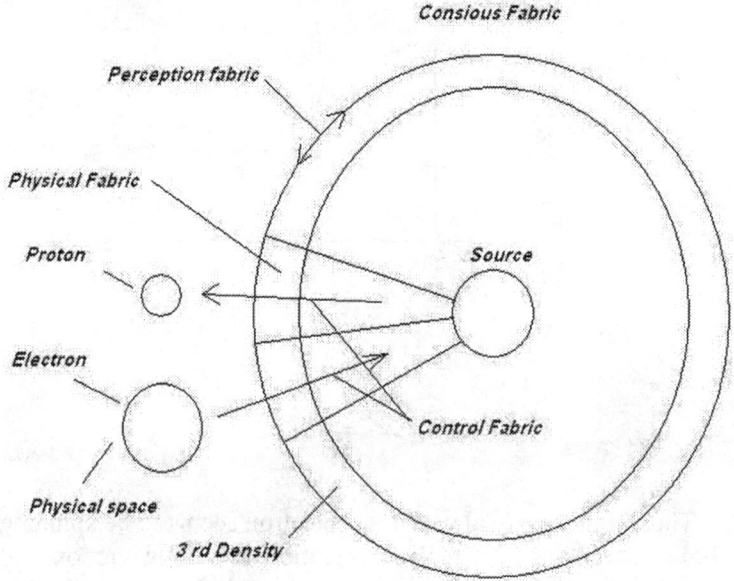

Using the control fabric [decision], perception spins around both the inflow and the outflow at light speed. One is Polarized with the inflow and one is polarized with the outflow. This creates the appearance of two particles in the perception fabric, Electron and Proton with reversed magnetic fields and reversed electric fields.

The tempic field appears to be located in the particles but is really located in the perception fabric looking back at the center of spin. As the only real forces are inflow or outflow and time flow rate, the perception fabric creates the perception of all other forces using light speed or unity spin around the real forces.

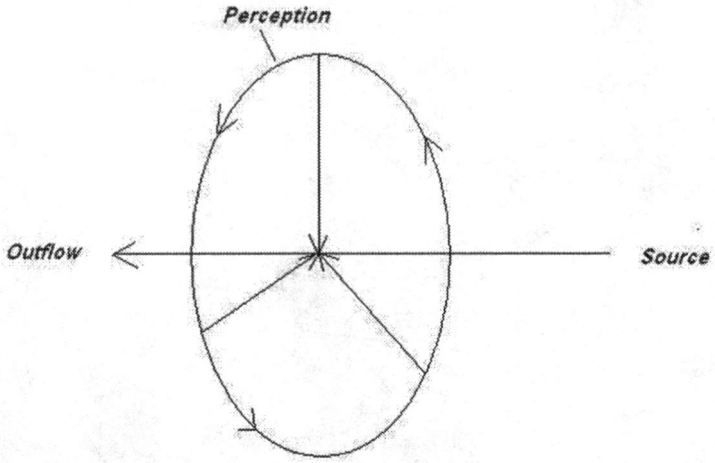

Thus in the physical world the Electron seems to be spinning at light speed, but it is really perception that is in motion and cannot be slowed or sped without invoking inflow or outflow to alter its density and it's distance from source. This is the function of the control vector.

This gives a measured level of insight as to why time flow rate is a function of the conscious layer and not the physical layer as now the Electrons spin velocity is identified as the velocity of consciousness observing itself at the smallest level of physical existence as it accelerates towards source, and Proton as it decelerates from source.

A finite piece of God Consciousness moving outwards from Source, contains both control and perception operating in quadrature. As the consciousness unit expands, it's "perception" is expanded along a conical shape as the area of a circle and thus opens away from source with the control fabric vector operation, setting in quadrature, acting to move it away from source. Along this outer side of the perception fabric the distance around the circle is greater then along the inner side and light speed would be moving slower, and this is the Proton the slower moving particle as perceived by consciousness on the physical side. This explains why gravity is really pulling away from source and slowing the time frame all at once, the Proton appears as the heavier particle.

The Rain Maker Device

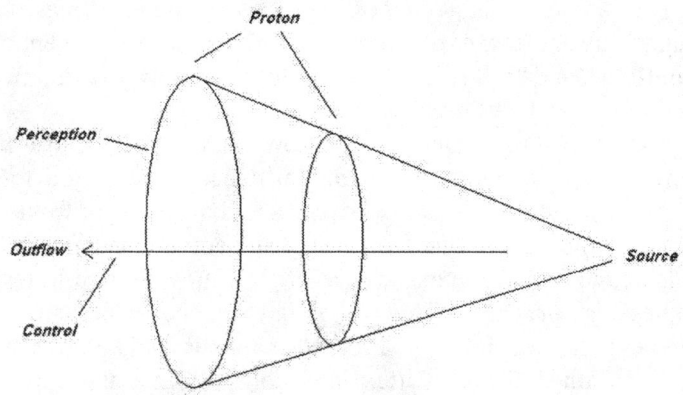

Along the "inflow" side the consciousness unit experiences the opposite and now light speed accelerates as it moves back towards source. This particle seems to contain a pull towards expanded awareness and higher time flow rate, contains more speed and less weight, the Electron.

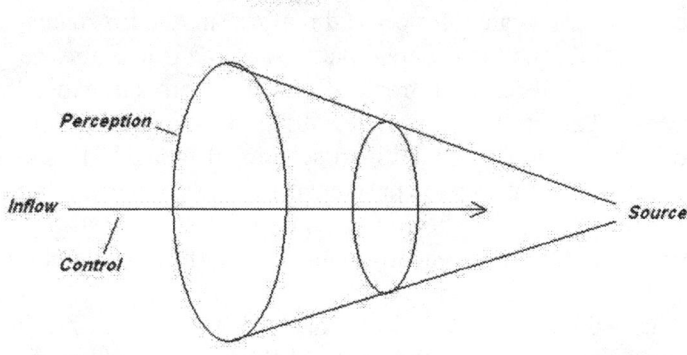

As consciousness moves away from source through the Proton, perception [light speed] is slowed, mass and gravity increase, and awareness may perceive the single point consciousness in "space" as the sensation of dragging time, "inflow". As consciousness reverses and moves back towards source awareness it is perceived as a lower gravity, increased or expanded awareness, expansion or "outflow". These perceptions of "outflow" and "inflow" in physical form are thus reversed from actual inflow and outflow from source.

Anyone who has faced the "inflow" and traveled into it has discovered that it holds the maximum distance from God force possible and yet it pulls ones consciousness down away from the God force into a place that feels heavy and compressed and could be likened to hell or the experience of slow moving hard to resist compressing pressure. This is found in the experience of awareness moving into the Protons magnetic field vortex and found at Cathedral rock vortex in Sedona Arizona and explains why all the homes in that area are up for sale. However since this is the path of consciousness entering this density from Source, it is the best place to go to merge with Source consciousness and begin to understand how the universe really works. It is the true "outflow" from Source.

Thus once again we are reminded of the spiritual truth that vocabulary is a function of the level of awareness consciousness is speaking from.

Words have different meanings depending on where consciousness is located. Inflow and outflow are the first example of this principle and the apparent paradox which is in reality not a paradox at all if viewed from all levels, from an expanded awareness. This is the crux of the differences of perception as found in the scientific and spiritual schools of thought. This also shows us where our human consciousness is really located in the physical as it perceives the world, and yet is also present on the inner side and able to equally fathom the spiritual nature of existence.

This would seem to indicate that now I may have reached zero point comprehension and accomplished my original goals of unification of both spiritual and scientific disciplines. The rest will now become moving back outwards to discover the methods of control and perception that may lead to harnessing the conscious fabric and removing mans need to live off the energy of other life forms, shifting it directly to drawing off Source energy.

Tempic Vector

The tempic vector is now seen as the deceleration of light speed [perception] as it enters our density through the conscious outflow. To perception it appears as a force of reversed tempic vector because perception itself is decelerating as it enters through the outflow. It is this force of perception being slowed, appearing to create perpetual motion in the atoms particles. Spin that never stops.

Light

"Light simply is."
This comprehension now gives an understanding of the basic photon as related to consciousness and light speed. Light does not travel through the perception fabric, it is the perception fabric.

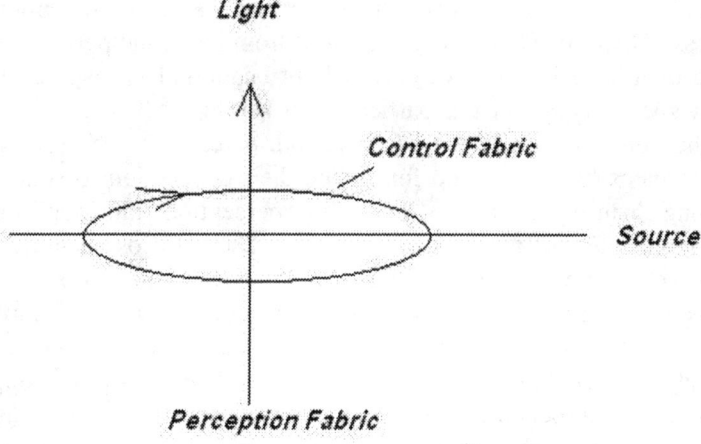

Matter is light speed consciousness, perception circling a control vector in space moving to or from source. Light is consciousness traveling through space along a linear perception vector with a looping control vector too short to take the conscious quantum out of this density. The control loop is altering lights velocity as it moves along the perception fabric. In physical space this would appear as a piece of a longitudinal wave and through it's frequency it appears to move faster then slower along its wavelength. Its bundle size is a quantum packet of fixed energy defined by planks constant and relative to frequency.

If we induce a reversal in the Control fabric that is too high in frequency to move out of the 3rd density, that is we reverse it before it has time to disappear and reappear as a quantum particle of matter, then we have created a photon, a small piece of the conscious fabric, that no longer reaches all the way back to source on each of it's reversals of inflow and outflow but instead stretches itself across space between where it was emitted and where it is reabsorbed. It disconnects from source, leaves the spiral path moving in and out released from its control fabric connection to source, and now begins to move along a relatively straight path as its spiral is released. Now it travels along the perception fabric alone. Moving, rather then in a perception loop around a polarized control vector, it becomes the entire control loop, its perception is spread out straight and it spreads perception across 3D space. Thus light as released from being the perception loop of a particle now becomes a light beam and propagates at light speed across all the particles still setting in 3 density. As "light simply is", in that it contains both perception and control, light beams have given up for a time their connection to source pulling them out of density, and their "perception spin loop" for the larger view of the physical universe. They still oscillate as a small control vector loop from and too source, however their perception loop is opened to a straight path. They would now contain both sides of the control loop in a single photon due merely to their higher frequency, and a broken perception loop popping out to become a line through space. Light takes the pure perception path only curving if it once again encounters an altered time flow rate strong enough to loop it around on itself and suck it back into an atom to energize the orbital path of Electrons or Protons.

Light beams now allow common perception to move outwards from source and establish "form" in a density.

The Rain Maker Device

They allow a new perception layer to form based on how the physical world appears from inside it that can disconnect from Source for a time. Without light, perception could not produce form inside a density. Light brings common reference to "multiplicity" and spreads perception across the physical universe. Indeed this interaction may be the roots of life as we know it as a Consciousness remaining separate from Source for the period we call a lifetime. An image of Source, for a time disconnected. Thus higher levels of consciousness are able to exist as separate from Source with smaller awareness, the human consciousness is a example.

Light travels in a curve through the universe:

As we set the perception fabric [light] free to move through the universe, it is actually becomes the conscious fabric. Since the time flow rate slows as we move away from source light without a control vector connection tends to curve away from source. Gravity is seen to bend light towards mass as it passes, thus gravity is pulling away from source. Since gravity is the only force we have observed to bend lights path it lies at a deeper level then magnetism and must be a function of the conscious fabric also.

Density

Visible Light does not cross a density threshold but is pulled back into it. Thus we do not see other densities, they are invisible.

Light by its very nature, a small particle quantum of looping control vector energy, cannot escape the velocity of its control loop on either end.

Thus light does not normally cross between densities. I would expect from this that no machine attempting to use Light could produce a density effect.

Density crossing is inherent within matter, because the tornado of particles reaches across all densities and back to source. So within matter lies the possibility of Over Unity through manipulation of its root connection with Source. Within matter lies the path back inwards to Source energy.

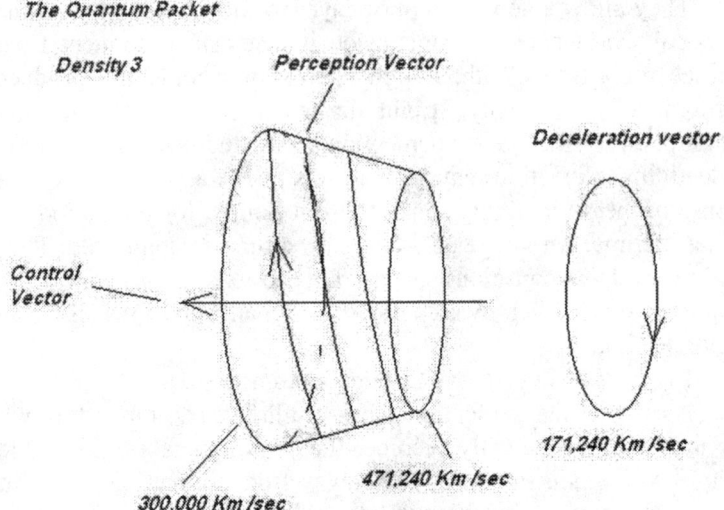

Within the quantum packet of matter we see the perception vector spiraling like a tornado expanding and slowing as it moves away from source, then contracting inwards and accelerating as it moves back into source. This happens around the control vector which is determining the direction. As the control vector moves out from source with perception spinning around it at light speed, light speed slows. When it reaches the inner border of our density it becomes visible to light now operating in its range. The quantum packet seems to "wink" on. As it crosses our density its velocity drops, and as it leaves it once again "winks" off. This is the nature of atomic particles at the quantum level, they simply wink on and off. It is during the on time that all the forces are present to interact, however the on time is not fixed as to light speed but contains a gradient. During this time the particle can interact with 3 forces in what we perceive as physical space. Inertial momentum, Electric field, and magnetic field, as well as photon emition.

A particle appearing to slow from 471,240 Km /sec to 300,000 Km /sec would be the result for an outflow particle and would appear to impart a dragging perception as it passes our 3rd density perception [life in 3rd density].

The Rain Maker Device

As to light speed and perception we generally only perceive light as it approaches us and never as it leaves us. We observed this in the magnetic field as tempic vectors moving away from one another have no effect. Perception setting inside a density will see the particle along its entry points but not along its exit points. Since light photons form during the interactions of matter quantum's acting on one another, they will tend to form near the center of the density setting up small control loops somewhere across the two density thresholds.

The illustration we see pictured above represents the half of a Proton we see as it approaches us, the half leaving center becomes invisible to any awareness at the center. Thus we see the higher velocity side of a Proton slowing, and we see the lower velocity half of an Electron accelerating. This would lead to the conclusion that Electrons generally probably produce photons centered around the lower to central side of our density and Protons produce photons on the upper side.

This divides the density in half and splits the central operating range for the Proton and Electron into dual velocity photon layers.

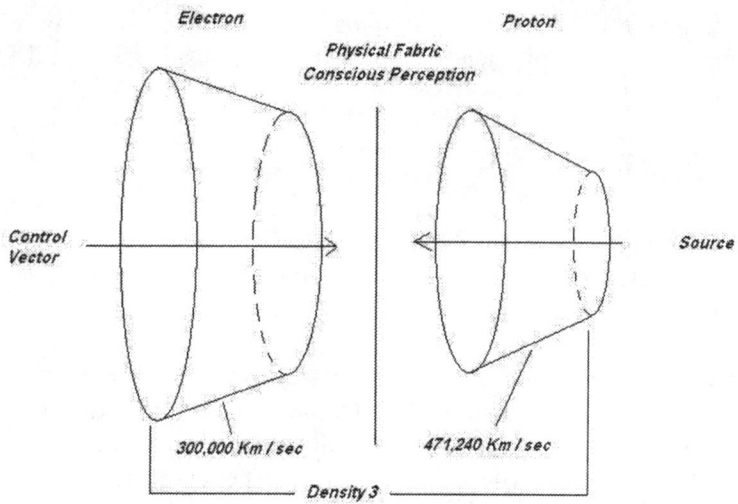

We have now split the density into two layers with different light speed photons, one centered around Electrons and one centered around Protons during the time they are visible. If the control vector loops are confined to less then half the density velocity spread, then the two types of photons will become invisible to one another. This describes what has been observed by Tesla in the T waves, as to Radio and visible light waves. A layer of the medium that propagates photon energy that does not normally interact with the Electrons magnetic field. The Protons photon exchange would then become invisible to normal light, however it would be lying on the higher velocity side of this density and thus may be the source of a higher energy as it is entering from Source directly.

So the actual density thresholds for 3rd density are probably somewhere above and below the two velocities given for photon emitions. It may be possible to find them using planks constant to reference the energy found in a photon of the visible light spectrum???

It would also appear that the human perception, physical eyes, are pointed at observing reality pointing away from Source towards the Electron layer and away from the Proton layer. The auric field thus seems to lie the other direction while seeming to appear outside of physical matter, it truly lies inside towards the higher energy path.

Definitions

Tempic - motion which is the basic parameter necessary to create the sensation of time flow and thus time is rooted in motion, motion is rooted in light speed.

Tempic field - the resultant field created from a quantum motion moving through 3 space with 3 tempic vectors of motion and results in a packet of perceived time flow rate which is the sum of all vector motions. The quantum CU is experienced as a single unit relative to other units of consciousness. The tempic field is the sum of all quantum CU's within its field of awareness. The tempic field is altered by all motion affecting the place where it is measured.

Quantum CU - The smallest unit of God consciousness perceived containing both a perception and control vector motion. As awareness expands outwards through perception it's sensation of time flow rate becomes a merging of all the quantum CU's within its field.

Matter - The result of quantum CU's entering and leaving a density, matter connects with Source on each loop through inflow and outflow on the control fabric.

Life - Consciousness disconnected from Source for a time and thus able to forget, while it perceives matter seemingly from outside.

Physical fabric - What life sees from the Source disconnected side of a density when it looks back at the Source connected side [matter].

Conscious fabric - If life chooses to connect with the inflow and outflow consciously it may move inwards to experience the Conscious fabric. The auric fields are discovered and the Higher self. The Conscious fabric is found to be more real then the physical fabric however perception is different and the tempic field is reversed. Inflow and outflow are perceived opposite.

T waves The "Tesla wave" propagates at a velocity of approximately 471,240 Km /sec as estimated by Tesla from human perception as probably no O-scopes were available for him to make an accurate measurement.

References

[Wilbert Smith]
I have listed the fabrics and quadrature aspects of each for a reference:

Space Fabric:
Length
Area
Volume

Field Fabric:
Tempic Field [Change / Gradient / Spin]
Electric Field [Divergence]
Magnetic Field [Curl / Deviation of Reality]

Control fabric:
Randomness Orientation
Decision Free Will
Ordered Sequence - Specific Arrangement

Perception fabric:
Form [Boundary of Reality]
Multiplicity
Assembly [Purposeful Structure, Animate / Inanimate]

Definitions

Source - God force, source of all motion and energy in the universe.

Conscious Fabric - Consisting of the Control and Perception Fabrics, the conscious fabric sets between the Physical fabric and [Source] energy powering the universe through perpetual motion and allowing awareness to perceive all the universe. The Astral and Mental planes have been identified with this function by Spiritualists. Astral brings the outflow moving awareness away from God and the Mental brings clarity and unity of intention and the path back to God, forming the conscious loop. The conscious fabric contains the source of the "tempic vector" and time flow rate.

The Rain Maker Device

Control Fabric - The path [Source] energy takes through the conscious fabric to produce intelligent motion in the physical plane. It is considered a real force and the source of all motion in the physical. The important math is that of "time flow rate" which is a function of distance from Source or center, for a density.

Perception Fabric - The path perception takes moving from the physical back to Soul which lives in the conscious fabric in one of two states, Life or Awakening.

Time Flow Rate - Light speed is not fixed, it is a function of the distance the "control fabric" sets from Source [Center of the conscious fabric].

Density - Stratified time flow rates create many densities along the "conscious fabric" where physical realms may exist. Moving between densities may cause a consciousness shift as well as an altered light speed constant to emerge.

Tempic Field - The tempic field defines the time flow rate and the tempic vector falls off as a function of linear distance in the physical fabric.

Prime Directive - The primal intention lying at unity is one and only one "intention" the intention to create division and conflict in the fabric of consciousness, and thus all the structures of the fabrics as they expand outwards, slow the time frames of awareness.

More on Creation of the Physical Fabric

The physical fabric sets along the conscious fabric at a given distance from Source where light speed is constant or perceived at one velocity. It takes the form of a sphere around source and is one of many other physical fabrics located at different distances were lightspeed is different. It is seen as the two dimensional area covering a sphere like an onion skin on the conscious fabric layer. It is a part of the conscious fabric.

As to light speed and perception we generally only perceive light as it approaches us and never as it leaves us. We observed this in the magnetic field as tempic vectors moving away from one another have no effect. Perception setting inside a density will see the particle along its entry points but not along its exit points. Since light photons form during the interactions of matter quantum's acting on one another, they will tend to form near the center of the density setting up small control loops somewhere across the two density thresholds.

The illustration we see pictured previously represents the half of a Proton we see as it approaches us, the half leaving center becomes invisible to any awareness at the center. Thus we see the higher velocity side of a Proton slowing, and we see the lower velocity half of an Electron accelerating. This would lead to the conclusion that Electrons generally probably produce photons centered around the lower to central side of our density and Protons produce photons on the upper side. This divides the density in half and splits the central operating range for the Proton and Electron into dual velocity photon layers around the physical fabric.

We have now split the density into two layers with different light speed photons, one centered around Electrons and one centered around Protons during the time they are visible and interactive. If the control vector loops are confined to less then half the density velocity spread, then the two types of photons will become invisible to one another. This describes what has been observed by Tesla in the T waves, as to Radio and visible light waves. A layer of the medium that propagates photon energy that does not normally interact with the Electrons magnetic field. The Protons photon exchange would then become invisible to normal light, however it would be lying on the higher velocity side of this density and thus may be the source of a higher energy as it is entering from Source directly.

So the actual density thresholds for 3rd density are probably somewhere above and below the two velocities estimated for photon emitions. It may be possible to find them using planks constant to reference the energy found in a photon of the visible light spectrum???

It would also appear that the human perception, physical eyes, are pointed at observing reality pointing away from Source towards the Electron layer and away from the Proton layer.

The Subtle Planes

No model of creation would be complete with out both halves of the God force, one flow moving towards the physical and one moving away.

As nature does not waste any of its parts the following diagram pictures the sides of the Electron and Proton vortexes that are not visible to physical existence. It is only consciousness itself that may move through these and experience them using awareness. Thus the auric layers lie only where consciousness may travel riding the control vector where it has no direct interaction with physical matter as it is moving away.

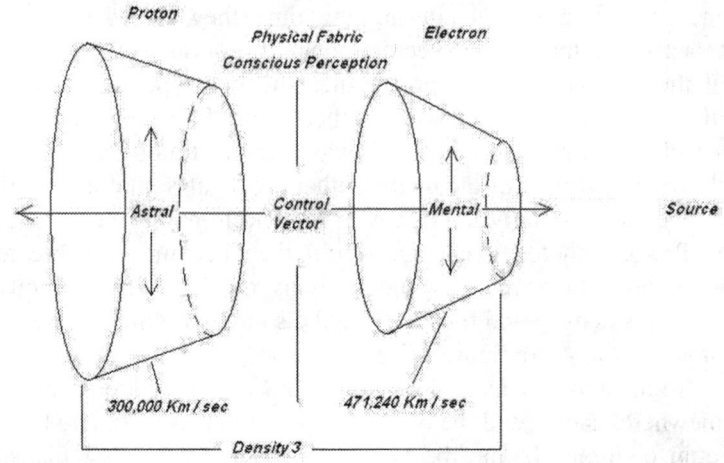

The human consciousness may slide along the control vector to a point where the perception vector of matter is now approaching it where it can be observed from outside the physical fabric. This has been labeled the Astral tube in Yoga.

When human consciousness encounters conflict it moves towards the control fabric and makes a choice to turn either towards the Astral or the Mental to deal with the conflict. The two choices lead to different outcomes and different actions. Astral interaction leads to division of intention and scattering, and mental interaction leads to resolving conflict and drawing together to a higher vibratory rate. Human Consciousness must experience both before it can discern the difference. Both have a function in our human consciousness and the evolution of our planet. The primal directive as observed in the Bible to fill the earth requires the scattering of the Astral plane function. Thus up until now Religion has been preoccupied with Emotion to fulfill this primal directive.

Neutron

One can now see from the above diagrams how a Neutron would physically allow the Electron to overlap the Proton and why they both have different sizes and slightly different density factors.

The Rain Maker Device

With all the forces present in the physical fabric the Neutron represents the greatest achieved connection of attractive forces to physical perception. However the conscious perception does not agree and finds itself split in two.

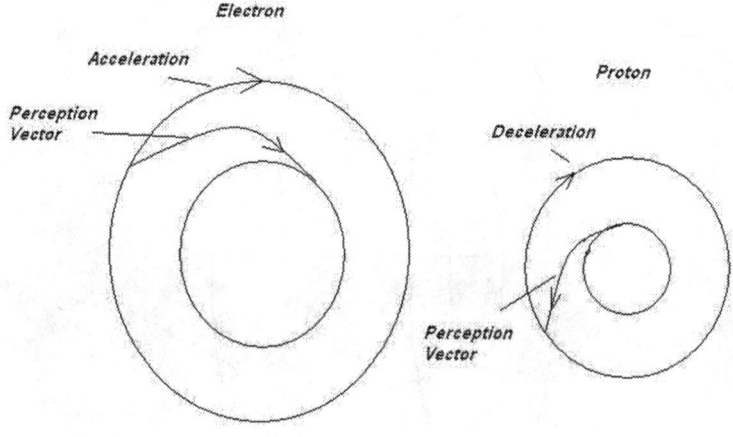

As viewed from the Physical Fabric

The graphic above shows us standing in the physical fabric looking out at both the Proton and Electron turned side by side.

The Protons perception vector is experiencing a deceleration creating a reverse tempic vector in the physical plane opposing its direction of motion.

The Electron is experiencing an acceleration of its perception fabric but the spin is the same direction on the conscious fabric as perception is moving.

The Protons tempic vector as viewed from the physical plane is reverse of its motion acting against its light speed velocity. The Electrons tempic vector in the physical plane is observed as the same direction of its perception vector acting with its light speed velocity. Thus the two appear to be spinning the same directions from the physical plane aware perspective. It is really the perception vectors that are moving in opposition as we view them from the physical plane.

We now overlap the two within our perception and it appears that tempic vectors are moving in the same direction.

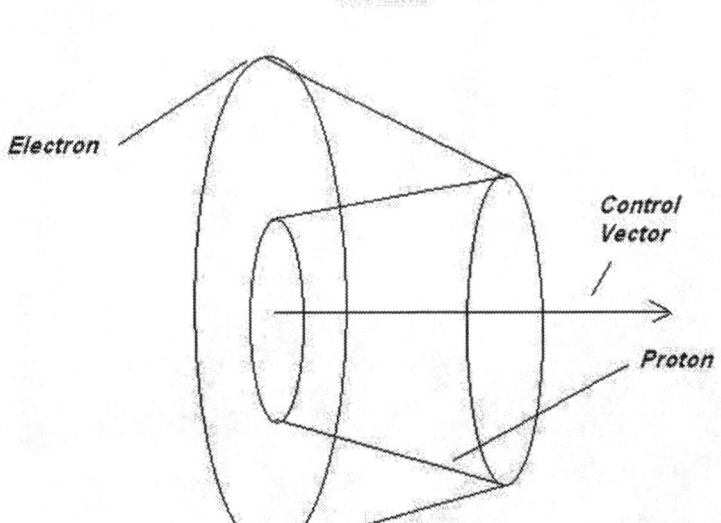

We see that within the Neutron is the overlap of control vectors moving through one another in parallel, but perception vectors are now in opposition.

The appearance in the physical plane however is different, and tempic vectors appear to be moving in parallel. This is because the tempic vectors of the physical plane are an observation of how they affect light speed, slowing it or speeding it.

The control vector in the Electron is moving towards Source and into the physical fabric, the control vector in the Proton is moving away from Source into the physical fabric, and thus both are moving into the physical plane and can be perceived as well as interact with one another.

In this configuration now in physical space the tempic vectors are moving in parallel, acceleration vector is now the same direction as deceleration vector, and thus attract, yet in the conscious fabric perception vectors are moving in opposition and are in reality expanding space. By overlapping the control vectors in parallel we allow the two particles to alter one another's shared size and density, we have combined the reality of both by over half. Thus Neutron moves to the center of density 3 setting between Proton and Electron and both alter their size accordingly.

Neutron is now found located exactly at the center of density 3 with altered Electron around altered Proton held together by the acceleration / deceleration vector forces aligned. The control vectors fall into attraction moving in parallel, the tempic vectors fall into attraction, and the perception vectors fall into opposition. It is this opposition of the perception vectors that achieve stability and why matter does not simply neutralize itself and disappear from reality.

This process is the result of consciousness spinning perception against itself, light speed set in opposition by intention. While in the physical we observe all forces in attraction.

Electron now is seen in a lower density with expanded space so Electrons interaction become more distant and it is found in orbits further out when exhibiting equal force. Protons are found closer in and operate in contracted space due to their operating along the upper side of the 3rd density band.

Creation of Physical Space

Spherical Form

The above models are relative to show the interactions of density, the perception and control vectors, in the physical fabric however this might lead one to believe that Protons and Electrons would now appear as flat circles in physical space rather then spheres, or at the least appear as looking down the vortex of a tornado. This is in fact how Charge functions. Indeed these forms are what is pictured in intuitive physics models [occult chemistry] as observed from the conscious fabrics. However we know from classical physics that the forces associated with Electrons operate in a spherical form in space, that is voltage is a spherical force as is gravity and yet actually operate as distance squared forces. Across the physical plane of space these forces are perceived as spherical and contain a magnetic field as they appear to spin the shape of a donut. Thus we now move out one level of motion and begin to observe the perception fabric laying inside density 3 that has disconnected with source as light along the perception fabric, the orbital motion and the quantum appearance of matter viewed from physical space.

Space

We must now develop a model of Space to make rational sense of the physical fabric in the only form possible.

The control fabric pushing out from Source could only take on one shape, spherical.

Perception thus spreads in all directions and may choose two paths along one surface of a density where light speed is constant. Perception can circle as matter, or it can move along the curve of the sphere as light. Lights velocity is only constant along one layer of the sphere. Thus given enough time light will travel all the way around the sphere of a given density no matter which way we send it because light is locked into one velocity for a given density.

Density becomes an area of a sphere's layers, like an onion. Physical fabric matter becomes perceptible only in the center of the layer and conscious perception becomes located only on one surface of the conscious layer known as the physical fabric. This layer is a perception looking only at perception vectors that are approaching it, as they pass center they disappear. We have created 2 dimensional space around a sphere where light speed is constant, and this is the only place in our conscious universe where light speed would remain constant.

Just under this layer we created a second layer where light speed is slightly higher.

It is only the magnetic field that can span both layers to perceive an interaction between the two and make use of the difference of density to use as a power source. When we look at matter we see only the 3rd density surface of it, so how do we get a 3 dimensional space?

The conscious layer is filled with light, and scattered pieces of God consciousness spreading out.

One layer of consciousness, as a density, now consciously inverts its perception and becomes the outer rim of space at one density.

Space is now the shape of a sphere operating from one layer of the conscious layer the physical fabric all with relative position in 3 space. Physical space is only a perception of the physical fabric in the shape of a sphere forming 3 space as we perceive it.

Everything that appears to exist as matter in physical space is the result of not less then two physical fabric conscious projections intersecting. Thus we can now give 3 dimensions to a space that appears to allow light to travel all at the same speed within it.

Life is but a dream in the sense we think of it and the Spiritual community should be very satisfied with this new view of the universe.

The scientist will resist the apparent complications from a model now demanding all physical space must be mapped into the conscious physical fabric spherical shell.

If we follow Wilbert's Suggestions, half of the perception of physical space would fall inside and half outside the physical fabric along the conscious layer. Just how we project images of consciousness outside and inside the sphere would be the next trick of consciousness operating from within the physical fabric.

Whether I have just succeeded at defining the universe, or the space of an atom is not yet totally clear however, but I suspect there may be a little of each found in this model of space. Whether the aura of an atom contains its physical fabric or the aura of the universe contains it may indeed not really mater. In the end we have observed a model at last that would seem to unify the operation of both.

Electric Force

Electric Force is now perceived as interacting spherically in space although we also know that it truly interacts as an area or distance squared force. We have identified the Electric vector as none less then the physical fabrics tempic spin around the control vector moving towards or away from source, the Proton volt is the distance between the perception circle where it enters density 3 and where it meets the physical fabric. Thus negative charged electrons are source pulling towards the physical fabric, and positive charged Protons are Source pushing towards the physical fabric. Why then do they attract? It is in reality the tempic vectors of acceleration and deceleration that attract, and not the perception vectors which repel.

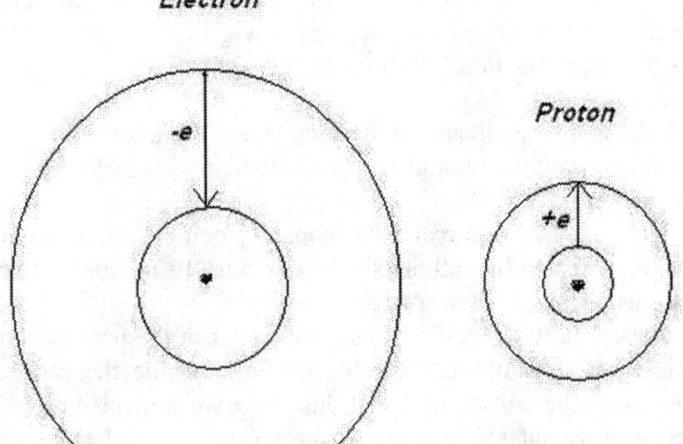

We saw from the Neutron model that as they come closer they actually expand the conscious fabric rather then destroying it.

The resultant tempic vectors as observed in the physical fabric attract one another and Proton appears to be spinning the opposite direction it is truly spinning.

This is why the electric force appears as an area and a distance squared force from any angle of perception we choose in 3 space.

The Rain Maker Device

3 space exists as a perception of the perception fabric trapped in density 3 and not as seen by the quantum particles or from source perspective.

The electric force thus always appears as the area of a circle which it truly is. As both the acceleration of the Electron and the deceleration of the Proton match as to the alteration of density along the control vector both forces appear as equal but with reversed charge and altered size. One is entering the physical fabric from a higher density [Proton] and one from a lower density [Electron]. Both contain the same length of the control vector so charge is equal.

From 3 space then the correct vector model of Electric force is the tempic vector spinning around the rotating voltage vector to form an area and this agrees with our notion that the Electron contains infinite spin and force that can not be depleted or exhausted.

Formation of the Control Loop - Precession - and the Magnetic Field

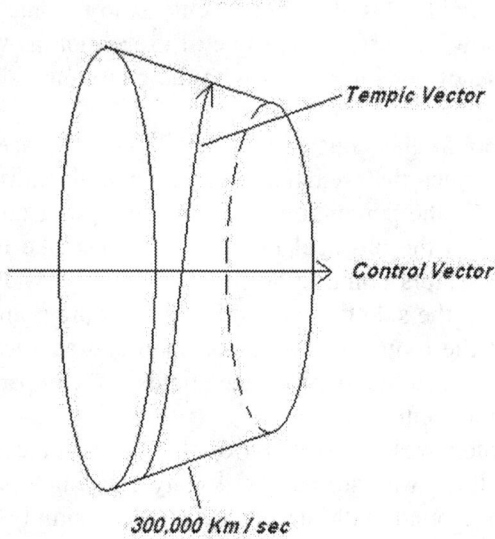

We have proposed that the photon is a loop in the control fabric as it moves through one half of the density around the perceptible half of an Electron or Proton.

What happens now if we create larger control loops that reach a little further out such that the perception fabric is crossing the density barriers, and the photons actually begin to interact between other levels? That is Proton photons now interact with Electron photons.

As light has already been associated with the magnetic field and an EM operation the next step is rather interesting.

A control vector with a loop large enough to span both electron and proton densities may actually loop around the entire density spanning both the visible and invisible halves of both Proton and Electron. The loop would carry the perception fabric along with it as the spiral spinning vector yet be perceived differently as it crosses the physical fabric.

The loop as we follow the tempic vector of deceleration and acceleration would appear to precess due to its slight curve forwards. That is if we fix our eyes on the tempic vector the control vector now appears to precess around altering its angle slightly, and here we find the motion of precession appearing in the magnetic field. This also gives a clue as to the magnetic field which we know can affect the angle of precession as well as the photon frequency for interacting with the conscious fabric of 3rd density.

The loop in the control fabric will now be seen able to produce an effect that remains locked on both ends still into source but rolls the perception fabric around into a donut shape. As it appears in the physical fabric it will look like two circles with voltage vectors pointing 90 degrees in at us. And the North pole will be on the side that all tempic vectors are pointing along the inside of the loop. Thus a Proton loop and an Electron loop will have reversed magnetic polar fields as compared to the direction of the control vector.

If a photon was released it would open the perception loop and extend all the way around the density carrying a piece of the control loop around with it, however if we simply delay the transition through magnetic attraction of Proton and Electron [repelling perception loops] the perception loop would now instead expand outwards and balloon up larger then the original entry point. Here we observe the interaction of the magnetic tempic vectors found in the vector model of magnetism. Looping just an Electron control loop would create an astral photon on the visible side and a mental photon on the invisible side. Looping a Proton would create a mental photon on the visible side and an Astral photon on the invisible side.

Looping an Electron and a Proton into one control loop would create a magnetic field that spans both densities and both conscious planes of awareness yet its ends would remain connected to source. Thus we see magnets offering much more then meets the eye. LOL! Atoms are now seen as magnetic fields crossing density, both sides of our 3rd density anyway.

Thus combining Proton magnetic fields with Electron magnetic field into interaction is seen as causing interaction in both photon mediums as well as both astral and mental planes of awareness. And this is what we see introduced in the Smith coils as I have been working with them.

From this observation it may be deduced that as the tempic acceleration or deceleration of the perception vector that produces the voltage vector, it is the perception vector itself that is producing the magnetic field. One is always in opposition between Electron and Proton and thus has been previously observed extensively.

Reflection on Creation

Mankind is now perceived as a two dimensional consciousness in a three dimensional world of physical space.

We are existing within the outer half of the physical fabric within our perception, the Electron lower density world, where light is moving at 300,000 Km /sec, and in the Astral existence lying within the Protons lower density half.

If we are now to reach a multi dimension consciousness and cross into the inner half of our density we must learn of the higher density side the world of Protons with photons propagating at a higher velocity. This will give us access to a higher conscious perception [the mental plane], and this is what I have been experiencing as I have used the Smith coil to slide inwards. Although our physical form seems to be trapped in two dimensions of the 3rd densities physical fabric for now, our awareness is not. Many have already experienced the greater depth of crossing the density boundaries along the control fabric via the astral tube and touched higher consciousness. Since the perceptions are mostly reversed it has been hard to realize further technology from these crossings. I believe it is only through the careful application of perception and observations that we can make the Neutrons connection and begin to work with two densities as the atom does, accessing the higher energy side of matter that is already present as well as the higher conscious level that comes with it.

The study of diamagnetics now appears to become far more critical, as the Neutron appears to be the crossing point between the two levels of perception.

Whether opening to the Protons layer will unleash a higher form of energy is yet to be observed. Thus reaching for the Proton layer will at the least bring a new level of awareness as our first step towards density travel and a higher consciousness.

Regardless of how the conscious physical fabric does this projection into 3 Space we have at least identified the real forces, how they interact, and why.

The Two Spin Universe

One of the other events we observe in physics is the nature of the electron to produce a 1 spin up and 1 spin down effect. The two vectors lie at an angle to one another. This would suggest a more complex universe where control vectors may intersect the physical fabric at various angles. As I can think of no reason why control vectors must move straight this makes sense. As the control vectors now take the shape of a donut they cross the light speed thresholds [densities] at various angles, perception spinning around them, releasing variously polarized light photons.

The Rain Maker Device

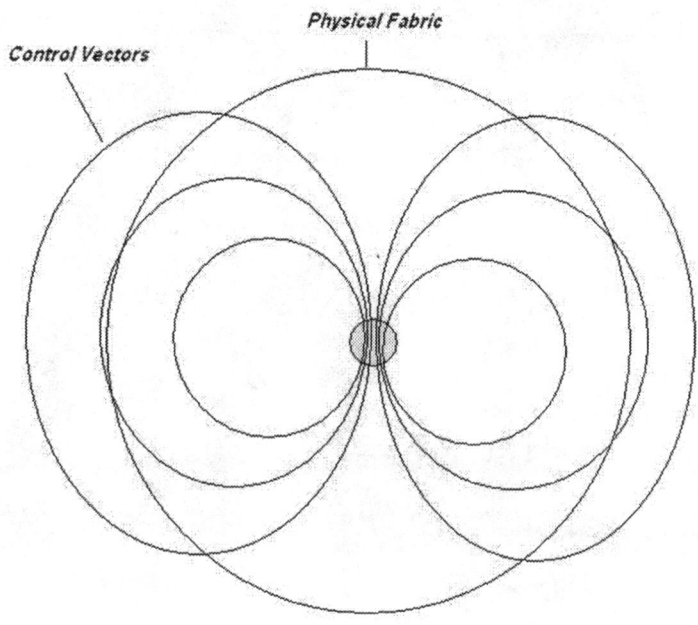

Thus as we move along the spherical physical fabric shell in the conscious plane we will now have access to forces pointing all angles possible as they cross the fabric. As we see the intersections aligning we recall that the perception vectors are approximately at 90 degrees to the control vectors. These angular force vectors can now be used to project physical space in any direction desired inside the sphere as conscious perceptions. However as at the top and bottom of the physical fabric intersection would all be vertical and along the sides would be very steep it may become almost impossible to project the perception of space into the poles and physical space may then be located more like we see it along a flattened cylinder near the middle. Interesting to note that now it becomes possible to project them outside the circle as well as inside it.

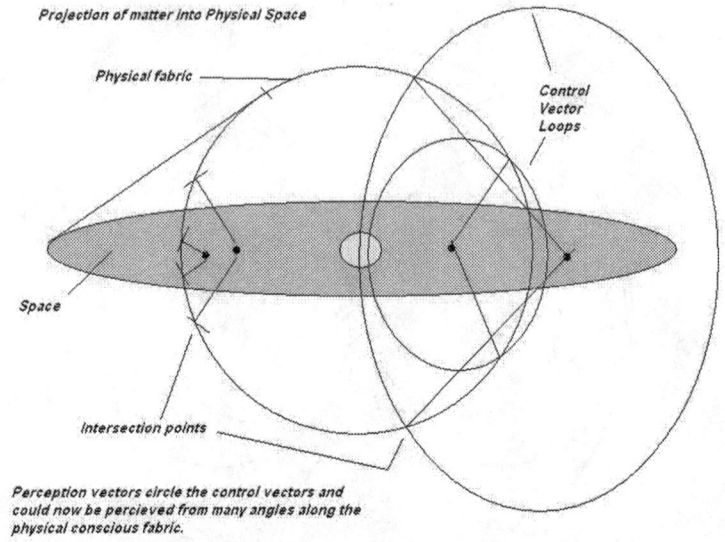

Definitions

Scalar - Having the quality of pressure or force.

Vector - Having the quality of direction, scalar force with direction allows motion.

Tempic - motion which is the basic parameter necessary to create the sensation of time flow and thus time is rooted in motion, motion is rooted in light speed.

Tempic field - the resultant field created from a quantum motion moving through 3 space with 3 tempic vectors of motion and results in a packet of perceived time flow rate which is the sum of all vector motions. The quantum CU is experienced as a single unit relative to other units of consciousness. The tempic field is the sum of all quantum CU's within its field of awareness. The tempic field is altered by all motion affecting the place where it is measured.

The Rain Maker Device

Quantum CU - The smallest unit of God consciousness perceived containing both a perception and control vector motion. As awareness expands outwards through perception it's sensation of time flow rate becomes a merging of all the quantum CU's within its field.

Matter - The result of quantum CU's entering and leaving a density, matter connects with Source on each loop through inflow and outflow on the control fabric.

Life - Consciousness disconnected from Source for a time and thus able to forget, while it perceives matter seemingly from outside.

Physical fabric - What life sees from the Source disconnected side of a density when it looks at incoming perception vectors [matter]. The physical fabric is a layer of perception that lies in the conscious fabric.

Conscious fabric - If life chooses to connect with the inflow and outflow consciously it may move along the control fabric to experience the Conscious fabric. The auric fields are discovered and the Higher self. The Conscious fabric is found to be more real then the physical fabric however perception is different and the tempic field is reversed. Inflow and outflow are perceived opposite.

Physical space - 3 dimensional space is the volume created in perception where light speed is constant along all its vectors of motion

T waves - The "Tesla wave" propagates at a velocity of approximately 471,240 Km /sec as estimated by Tesla from human perception as probably no O-scopes were available for him to make an accurate measurement

* * * * *

This section begins with Progressive theory [opened ended science], my continued seeking to comprehend and add to the "New Science" as started by Wilbert Smith as compared to the conventional models found today in physics. It then offers some experimental evidences as a record of the progress leading to observations of the interaction of the real world. It is a progressive process that changes often and many have contributed to the progress.

A Short Summary

The basic principles:

The diamagnetic field is the root of the atoms operation originating in the Neutron and stabilized all the orbitals using repulsion off both poles of the orbitals.

Gravity is the result of an altered time flow rate in matter as matter moves slower along the inside of a tempic gradient appearing always in a spherical shape. All atoms thus spinning inside a tempic gradient will be pulled towards the slower moving time flow rate but are actually turning slower along that side. Thus tempic vectors can explain all forces at their core operating level.

Torsion or "time flow rate" is coupled through a diamagnetic field.

The materials Copper Aluminum and Bismuth are magnetic at the Proton layer tightly coupled in the strong force area of the Nucleus.

The materials Iron and Nickel are magnetic at the Electron layer.

Tesla suggested the Time flow rate differential is around 1.5708 between T waves and Radio waves propagation velocities from direct experimental observation.

This suggests the tempic gradient appearing across this density.

The vortex generator:

The Rain Maker Device

Coupling a Proton magnetic material to an Electron magnetic material with a magnetic field allows the two time frames to interact as a function of magnetic field strength and time, through the B field which spans both "mediums". Using a scalar canceling coil to reduce the Electrons magnetic field it becomes a mirror reflecting the Protons magnetic field back inwards.

If this process is followed into a cylindrical form, a spherical field is set up altering the time flow rate of the sphere.

If we alter it the correct direction then gravity should be lowered.

The magnets polarity determines whether the field is a slower time frame or a higher one as it couples the Electron to the Proton in only two possible methods.

Protonics:

The study has led me to a new model of Protonics being derived through direct experiment, which is attempting to define the parameters of time flow rate and gravity. Also redefining the source of the diamagnetic field to be centered not in the Electron shell but the Nucleus.

The New Science:

Finally my last two files added are an extension of Wilbert Smiths Perception and Control fabrics which I refer to merely as the Conscious fabric and lead to a new model of Creation of the physical world, allowing mathematics to move through and into the Conscious fabric. This is perceived by me as my greatest realization to date, bringing a merging of the Spiritual and Scientific disciplines finally into a single model and a new comprehension of how time actually functions.

I just made some bismuth coils and discovered some things.

The very best coil is one made of Aluminum tubing 1 x .055 inches round. Cut the tubing with a pipe cutter then melt and pour it full of Bismuth using modeling clay lined with aluminum foil to create the form to hold the tube upright while it cools.

Wrap only one layer of wire 24 gauge insulated in the weave pattern back and forth down each side. Wrap it with a layer of electrical tape to keep the wires in place.

When this coil is lying on its side inside the iron ring it produces the strongest effects. You can short the wires together to cause the Proton side of the effect.

On the iron layer [the electron effect], you can find an old TV monitor from any old computer and carefully remove the ferrite ring from the picture tube. It splits into two sections if you remove the retaining clips. This is the ring setting inside the deflection coils, and care must be taken not to break it as you cut the wiring off it and tap off the glue and plastic parts.

Then place the retaining rings back on the two halves and you have the perfect ferrite ring as you see in the photos of Rain Maker.

The magnets need to be about 20 each 3/4 inch discs from K & J Magnetics on the bottom four sides of the ring, then the rest of the magnets can be smaller. Magnets need to cover the ring so that when a compass is passed around the ring one pole is always inside the ring and one pole is outside the ring all the way around it. No reversals of the magnetic field anywhere along the iron ring.

The coil on the outer ferrite ring is wound with two separate conductors so that one can be wired in reverse, this creates a scalar cancelling coil on the ferrite ring setting 90 degrees to the magnetic field if you want to connect a generator. Simply short all four wires together. I know this sounds strange but it works. The wires can then be taped so that the coil form holds the shape of the ring. This coil can easily be removed because the tape holds it in its form.

I was instructed to present this question.

What happens if we wind a magnetic cancelling coil on a magnet with a 300 year lifespan? Does it cancel the field?

Orientating the cancelling coil at 90 degrees to the magnetic field of the magnet puts it in alignment with the control vector for density crossing. This is an energy path that is never depleted and a true link with Source flow. The ballooning magnetic field is only a perception vector.

The Rain Maker Device

The generator without function generators, windings shorted, will respond to the conscious connection of the mind and the field pops out and is interactive because once established between the Bismuth and the iron it continues to flow.

I hope this helps anyone attempting to construct one of these.

If you decide to use an iron nipple instead of the ferrite ring the softer the iron the better. Rigid conduit fittings worked but iron pipe fittings work best.

The addition of the crystal ball is also very helpful in starting up the field and keeping a vibration present. It must be a true quartz and not glass. The field will move along its surface very nicely.

The first time I did this experiment I established a field that would not go away even after I moved the device. I have now realized that it was me feeding it all along mentally keeping the field present. I have since been able to build fields and then have them totally disappear by removing my fixation of mind from the area.

I know the goal is to come up with a device that works "dumb" so no conscious projections are necessary, however playing with this field gives one a new depth of energy and what it really is. All energy is linked to some perception vector, whether it is our human focus or another one.

Dave Lowrance

Magnetic Vortex Generator - Construction Detail

Aluminum Bismuth Scalar Coil

This coil is the heart of configuring the device for a Proton magnetic outflow or inflow.

There is much leeway in the coils, they can be custom sized to fit the device however try to keep the diameter under 1". 3/4" is optimal and bigger is not better. If a crystal ball will be used keep the coil shorter then the ferrite ring so the ball will sit on top.

Materials

Aluminum tube - 3/4" diameter cut to 1 1/2"
Bismuth shot BB's [enough to fill the tube about twice there will be some waste]
24 gauge copper or tinned copper hookup wire about 20'
Electrical tape grey or red as black is rather ugly
1 Pkg modeling clay

1 small cast iron melting pot
Aluminum foil
Cookie sheet or other small pan
1 Propane torch - [or stove top can be used burner on high 520 deg F]
1 hot pad mitten for pouring
Safety glasses

Procedure

The coil is made from an Aluminum tube 3/4" outside diameter 1 1/2" long. It is filled with Bismuth. Using a small pan cover the bottom with Aluminum foil. Set the tube upright and then seal it on the outside with a thick ring of modeling clay to contain the bismuth from escaping as it cools. The Bismuth is shot BB's that can be found at a reloading supply store. It is melted in a small cast iron melting pan that can be found at a kitchen supply store. It takes about 520 degrees F to melt the Bismuth, I used a propane torch. Bismuth expands as it cools so do not fill the tube completely. Take your time in the melting and slowly swirl the pan to get all the shot melted. There will be some scum at the top but as you pour the liquid bismuth the scum will be the last thing out so do not worry about it. Let the core cool well before touching it! The best coils will crystallize along the top of the coil as they cool and heating the metal longer and hotter will help this.

The Rain Maker Device

Next wind the coil:

Coil winding Detail

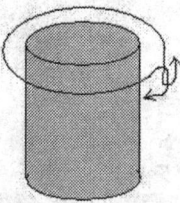

Start in the middle of the 20 foot wire. Loop the core at one end. Twist the wires around each other once so that wires trace back the way they came.

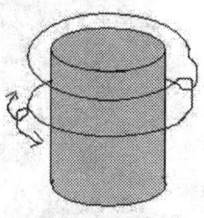

Repeat this on the opposite side each time down the coil keeping windings tight and as close together as possible.

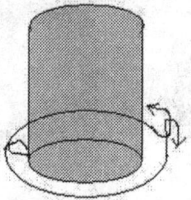

After the first layer is completed wrap the coil with one layer of electrical tape and then wrap another layer carefully up the coil.

This time though place the twists at 90 degrees to the underlying layer.

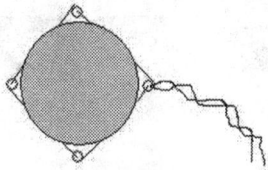

The finished coil should look like this now from the top. Carefully wind the two leads about 1 turn per 1/4 inch and cut at about 2'. A layer of electrical tape over the outer layer will keep the wires in place.

Ferrite Ring

The ferrite ring can be scavenged from an older computer monitor, the bigger the better. It is a slow job and care must be taken not to break the ferrite as the chalking is chipped off near the bottom. The ferrite ring sits inside the yoke on the back of the picture tube. If you do not know anything about TV sets and discharging the lethal voltages get someone who knows as the voltages present even in a powered down set can be lethal. Be sure to keep the metal clips that hold the two halves of the ferrite ring together. If a ferrite ring is not available a 2 1/4" or 2 1/2" diameter iron pipe nipple can be used about 2" long.

Magnets

K & J Magnetics

Many patterns can be used to set up the magnetic field however Neo magnets are recommended of sufficient quantities for a strong outflow to be achieved.

Here is my latest recommendation:

24 each type DC2 3/4" x 1/8" disc [placed around the bottom in six locations of 4 each stacked]

24 each type DA2 5/8" x 1/8" disc [Placed around the top in six locations of 4 each stacked]

Optional magnets can be added to these of a smaller 1/2" x 1/8" size to expand the field's width. D82 is a good choice.

If a large crystal ball is used this will expand the field envelope. A good magnet layout may cost around $60.

Use a compass to make sure magnets are placed correctly. North pole should be out on all the magnets and no reversals should appear anywhere along the ferrite ring. Depending the size of the ferrite ring it could take more or less magnet stacks. There will be a maximum limit of magnets before they start to push one another off.

Ferrite Coil

The Red coil around the ferrite ring is optional and offers a scalar interaction with the iron atoms. It is used to create an Electron inflow or outflow. It is made by laying red electrical tape turned sticky side out along the ferrite then wrapping a twin lead 24 gauge wire around it between the magnets as thick as desired then wrapping the tape over to hold the shape of the ferrite so the coil can be removed easily. To configure this coil for scalar operation the wires from each end of the coil are fed opposite directions for a magnetic canceling effect. As an option one end can be shorted and taped then the coil is fed from the other end.

Function Generator

A program is available on the Kosol Core Tech group that has been especially designed to drive the coil using a computer sound output.

If this is used an 8 ohm to 100 ohm resister must be placed in series with the coil to protect the sound card of the PC.

Other wise a function generator can be purchased to drive the coil. Ideally one can experiment with frequencies, however NMR resonance effects happen between 1 and 20 MHz for the metals present. Frequencies even as low as 8 Hz have been used, but I recommend the higher frequencies. As one works with the frequencies they normally tend to increase them over time.

Crystal Ball

The crystal ball pictured is a 125 mm Diameter Calcite sphere.

Quartz crystals have a natural electric vibration and any spheres can be used.

Dave L
and
Kosol Ouch

Conclusion

Welcome back,

I have put up some photos of my Iron pipe experiment per your request, but you have beat me to it I guess. I get a very good flow from the steel pipe rigid conduit, and what is even better you can get a size to fit much closer around the Bismuth slug and this creates an even stronger effect.

I highly recommend using the steel pipe to experiment with.

Wind a scalar coil in the center of it and then place magnets around upper and lower ends.

The long magnets seem to bring the energy out more for a bigger circulation area.

Since the diamagnetic field of Bismuth repels all magnetic fields, by surrounding it with a tight cylinder of field pointing inwards we end up creating repulsion from all directions, and this expands the time flow rate enough to create a sensation.

The latest paper brings a new model for "time flow rate."

I have found a certain beauty in this piece of work and a personal satisfaction.

Dave Lowrance

Physical Fabric Paradox

As we examine a circle in physical space we observe the first contradiction to the control fabric. For a tempic vector to travel in a circle it must turn more slowly on its inner side and thus time appears to move more slowly towards the center of the circle then along it's outer edge. This is the only way that a vector of motion can move in a circle and thus the first reversed perception is born, that center is dead and time is frozen at the center. This is how matter appears yet stands in complete conflict of how things work on the levels of the Control fabric. From this observation we must conclude that the operation of matter is reversed from the true operation of the universe and that Electrons physically setting outside the Nucleus are in fact a closer unity and thus time flow rate is higher as we move outwards in the physical plane. Thus a higher density is one with increased physical radius. Embracing a larger physical awareness brings unity in the control fabric and thus truly speeds the time flow rate. However perception of this is reversed in that as we actually move to a higher time flow rate we find a unity of intention and the physical universe appears to shrink from the perception fabric layer. This demonstrates that the Control fabric is the true reality and the physical is the secondary reality in opposition. As we observe inflow and outflow sensation this reversal is also true. While standing on Bell Rock in Sedona Arizona one is expanded in consciousness sensing physical out flow and yet simultaneously drawn inwards to God consciousness. While setting in the inflow of Cathedral rock one is draw physically inwards only to discover every intimate detail of the causal events of each and every conflict they have ever experienced within separateness from God consciousness and personal Karma.

As consciousness moves outwards to embrace all physical matter the time flow rate increases, and as consciousness moves inwards to the single point in physical space it becomes lost in the infinite flow of time as time flow rate decreases and even seems to halt.

The physical fabric lies in quadrature to the conscious fabric, however perception is reversed as to its operation. As Source divides itself outwards in the conscious fabric, it spreads itself out over all the physical realm moving inwards towards a physically stationary center frozen in time. This creates the illusion of entropy. As physical space expands, consciousness slows the time frame of a single point awareness. This is the nature of the tempic vector appearing oppositely in both fabrics.

The meditators have shown us the conscious fabric and the scientists have shown us the physical fabric and both sit forever in paradox and contradiction until an expanded awareness is available to explain and relate them both.

From the above we realize that gravity is a force pulling us physically inwards towards a slower time flow rate and was placed by intention to separate or divide consciousness into many forms and possibilities all separate in infinite variety, and thus we observe a universe of infinite variety located in small pieces of God awareness with a small single point awareness. To unite diverse galactic cultures is to move awareness back outwards and begins the reunification of God force itself inwards. This begins on the social level and culminates in the merging with God consciousness for all, but the formula is always the same, recognize conflict at all levels and resolve the diverse conflict of intentions.

Space was the first creation of consciousness to allow for the separation of individual conscious forms in opposition of intention.

Consciousness traversing space is the beginning of reunification and drawing back to Source. Thus the draw to move outwards physically and begin to explore the space that has separated all consciousness into separate pockets of [Source] dividing itself. As consciousness moves outwards in the physical it is moving inwards in the conscious fabric of perception and control. When all separate pieces of consciousness once again reaches the boundaries of the universe then all will become God and the universe will be done with its experience of itself.

For now, consciousness finds attraction in the perception of opposite control forces manifest as beauty, at perceiving itself in opposition of intentions in infinite variety. This is the prime directive of Source and nothing else, to cycle between the outflow and the inflow.

A machine to traverse density therefore would act to physically expand matter at its deepest level, the Neutron, turning the parallel physical forces into what appears to be physical opposition, which is in actuality unification of intention causing the true size of the universe to shrink along the conscious fabric. Only through physical expansion of the smallest level of awareness can we shrink the universe to a size that can be quickly traversed.

Physical Reality

There are only three forces, all else is derived from these interacting and appearing in alternating configurations of the appearance of opposition. They sit in quadrature. [Wilbert Smith]

Tempic
Electric
Magnetic

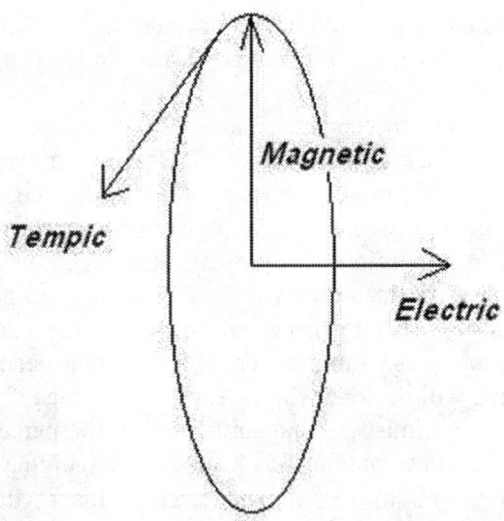

These three are further manifesting in three spatial configurations of interaction setting up a universe of force vectors in all directions and combinations that can manifest opposition appearing as attraction to awareness. The interaction of time vectors has been covered in the time vector model of magnetism previously and voltage and magnetic vectors can be referenced elsewhere, suffice to point out that the tempic vectors are in actuality operating in reverse of the conscious fabric.

Time vectors moving in parallel while appearing to form attraction in the physical are actually creating a slower moving time frame and pulling away from Source awareness causing an expansion of the universe of the control fabric. The resulting forces listed next are all derivatives of the ones above in combinations of alignment.

Strong force
Time flow rate
Diamagnetism
Gravity

The atom consists of three layers of these forces in three combinations.

Neutron layer
Proton layer
Electron layer

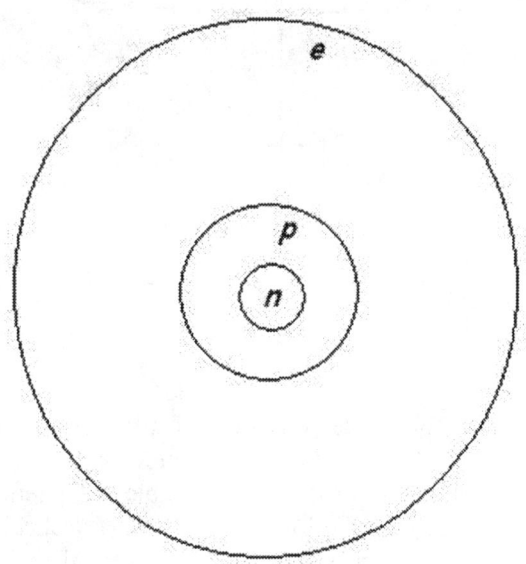

Using the tempic vector model of magnetism if we look at the interaction necessary in each layer we may gain insight.

Neutron Layer

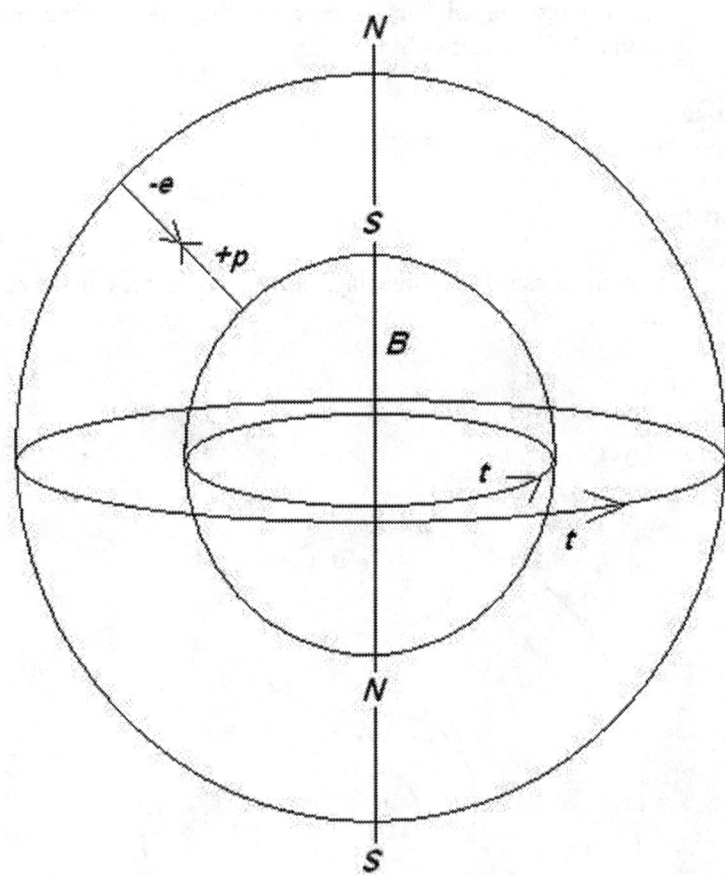

The Neutron is a composite consisting of a Proton setting inside an Electron and in this overlap all the forces between both fall into attraction at the closest range possible. Magnetically the Neutron is the equivalent of wrapping a scalar canceling coil, only instead the coils are spherical or cylindrical and one sits inside the other with reversed magnetic polarity. Also the Electric vector is not moving along the coils wires but pointing straight outwards/ inwards and this places the tempic [t] vectors in parallel moving at 90 degrees to the voltage vector as though it were moving through the coils wire instead of the electric current. Since it would appear that Voltage is setting 90 degrees to motion, the root source of power in our realm is torsion and not Electricity. Whatever is spinning the Proton and Electron is [Source] and this is the tempic vector. The prime mover. This is the 90 degree mystery force found in the spinning Copper cylinder experiment. As viewed from inside, looking out, Spin is one direction, it is magnetism and voltage that are reversed in the two particles and they are first formed in attraction. As Consciousness [control fabric] increases a particles radius by pushing it further out against this attraction the particle spins faster towards lightspeed.

This creates a time flow rate alteration or a tempic vector radiating from the Neutron much stronger then between the orbitals because of it's very short distance.

Thus the field resulting has no voltage vector imbalance, it is a scalar balanced voltage field, but has a magnetic field that can fluctuate with the tilt angle of an external magnetic field, and a time flow rate vector that drags or slows the time flow rate and this is the diamagnetic field as we observe it.

The Neutron setting outside the atom is seen as a weak Electron magnetic force, and this is all. It has no external Electric force interaction, however setting next to a Proton inside the atom it becomes the strongest force of nature we have identified, actually converting mass into energy. This is because it is one of the strongest creators of mass and thus has the slowest time flow rate due to the tempic vectors traveling parallel. The Neutron although appearing to be almost neutral is in reality the strongest particle of them all. The strong force is a result of two magnetic fields situated such that their tempic vectors run in parallel, their voltage vectors spherically are as close to a short as we can get, and their magnetic vectors are in attraction flowing directly through one another.

Due to the half or "reality overlap rule" offered by Wilbert Smith the diamagnetic field emerging has a stronger tempic and magnetic field then the outer orbital shells and we find when the Proton orbital tempic vector aligns also the three tempic vectors in parallel create the strong force at 137 times the forces found further out.

Normally the magnetic field shrinks inwards, and the Neutron remains with only a small negative magnetic moment matching the Electron, but as an external magnetic field is applied it interacts with both of Neutrons magnetic fields to push them out of balance drawing away some of the flux of one of the fields and reflecting the flux of the other, and a stronger force emerges to repel the external field. The harder we push them apart the stronger the opposing force we get back. This becomes the source of the diamagnetic field force.

Proton Layer

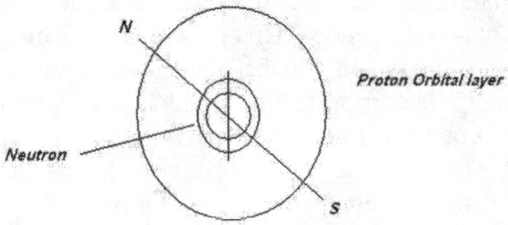

Neutrons interactive dual magnetic field system offers the maximum attraction for all vectors of forces in all three physical dimensions.

As flux is diverted outwards by the Proton field the Neutron alters it's magnetic output to always provide repulsion to it. The diamagnetic field is born here.

The Proton layer is setting very close to the Neutron and within its altered time flow rate, called the strong force area. Protons magnetic field is setting in some alignment with Neutrons to produce the diamagnetic field it floats on repelling both its magnetic poles. It's orbital tempic vector is aligned very close to Neutrons making three nearly parallel tempic vectors and attaining the strongest attraction possible 137 times stronger then the Electrons fields. Its particle spin field also aligns its tempic vector and adds to this force a 4rth tempic vector. However its magnetic field is free to expand and contract. The tilt angle of the Protons magnetic field sets the diamagnetic field strength and all the other atomic shells interact through this collection of connecting forces. The Proton shells voltage vector is not scalar cancelling within itself and radiates outwards spherically from the entire shell. Proton is setting in the strong force area and radiating positive electric potential as well as transferring the diamagnetic field outwards through it. In order for the Proton layer to speed the time flow rate it must push the Neutrons two magnetic fields apart separating the tempic vectors angles of motion. It is here we see [mass into energy] as the tempic vectors tilt away from a parallel path and the strongest time flow rate vectors are affected.

Electron Layer

The Electron layer is the lightest and fastest and its magnetic field, aligned with the Protons in one of two possible alignments, is expanding along all three tempic vectors of motion all repelling one another. The Electron magnetic vector is the strongest for EM interactions. Interaction with the Neutron layer will allow flux to expand to the highest time flow rate possible in any of the particles.

Diamagnetism

Diamagnetism is a force identified to repel both ends of a magnet. This is probably the most complex force encountered in the study of atomic motions and hardest to fathom as no one to date has created it using coils.

    This model is offered as one felt to be closer then conventional explanations offered today in classical theory that assume it comes from the Electron layer alone, failing to recognize the need for the force to be present inside the atom as the main force involved in it's basic structural operation. There is considerable conflict as to what keeps the Electron from crashing into the Nucleus, but classical physics has identified the force must be Electro magnetic as there is none other present to any level strong enough to accomplish this.

All forces in the atom are based on the three primary fields of force, tempic, electric, magnetic. [Wilbert Smith]

From the Protons outwards Magnetism and Electric attraction are the only significant forces in operation, therefore the atomic shells remain in their orbits due to magnetism, electric, or tempic vectors of motion and nothing else. [This is from modern physics although no one has offered a reasonable model to explain it.]

The nucleus of the atom must exhibit diamagnetism or the Electron shell of iron atoms would fold inwards on one side and shoot outwards on the other side.

The nucleus must be repelling both poles of the Electron shells magnetic field equally as well as both poles of the Proton layer which also falls into orbital shells.

Yet we know from NMR it is possible for the Proton layer to swing either into Electrons magnetic attraction or into repulsion based on its energy state from photon absorption. As it does, its motion or tempic vector reverses direction. The voltage vector will always align straight between them and can never shift expand or contract. So we end up with two configurations between Proton and Electron layers, one where magnetic B field vector opposes, and one where tempic vector [motion] opposes, while the voltage is a constant pull.

From the above model of a Neutron, what happens when we set it in an external magnetic field?

That is, how does the Proton layer affect the Neutron layer to create a diamagnetic interaction and establish the diamagnetic field?

We see the Neutron setting with two overlapping magnetic fields in attraction one inside the other. If left alone nearly all the flux may be expected to circulate between the Neutrons parts, only the tempic vector is moving outwards to create the strong force area. The Neutron however has a small negative magnetic

The Rain Maker Device

moment giving the Electron inside it dominance and thus a few flux lines still move outwards around it to interact with the Proton layer.

As the Proton shell interacts with this outer Neutron flux path in attraction the Electron flux is weakened or pulled outwards diverting it from its inner flow and the inner Protons stronger field expands outwards and repels the Proton outside because it is also in attraction to the Electron shell of the Neutron. The first or inner diamagnetic field is born forming between the Neutron and Proton layers. The Neutron has two vectors that can be expanded or decompressed through interaction. They are the magnetic flux field and the tempic or time gradient field. Both are expanded through repulsion of vectors in opposition.

Through a repulsive interaction the Proton shell has opened one tempic vector of motion to release some of the compressed field of the Neutron to expand outwards, and time flow rate increases for the nucleus which now becomes lighter then the sum of its parts and we see "mass into energy". We know that removing the Proton and Neutron from the atom increases their weight and the strong force disappears.

As the Electrons large magnetic field interacts with both inner layers, in one of two alignments with the Proton shell, pulling flux outwards through attraction from the Proton layer, and releasing tempic energy through repulsion, it now finds a balance of opposing and attracting fields to connect with no matter how it turns. Any flux drawn from the inner layers causes a shift all the way into the Neutron creating an opposing field to reflect back outwards which is stronger independent of the position of the Proton layers polarity to it. We find the second layer of the diamagnetic field between the Electron and Proton layer. The diamagnetic field moving outwards offers two variable vectors, a tempic and a magnetic one.

It would appear the Neutron must be the true source of the diamagnetic field, and the close distance of the Proton layer setting outside it, and this mechanism may be responsible for the alternating shells forming of both Protons and Electrons all maintaining diamagnetic fields to hold their orbital shell positions. The repelling force is all done with magnetic fields and tempic vectors of motion in three dimensions as voltage is a strictly spherical force appearing between only the Protons and Electrons.

Mass is located in the higher tempic compression areas of the atom at the core where tempic lines move in parallel and create attraction.

This model allows for the Proton layer to swing either direction around the Neutron layer as well as to the Electron layer, however the Proton layer would normally be setting in repulsion to the Neutrons dominant Electron field and we may expect it to be precessing at some angle where both Neutron fields are equalized. The tilt angle between Protons magnetic field and Neutrons diamagnetic field will determine the balance and all outer shells will interact through it causing it to regulate the Neutrons diamagnetic field.

As Neutron and Proton have about the same mass, precession would be present in both, but moving on opposite sides of the same axis.

The other possibility is that it could be located at 90 degrees to the field and precession is not along a common axis at all. This may be the case with Bismuth which has a strong diamagnetic field that is external to the atoms.

Diamagnetism is externally present in all non magnetic atoms, however in Bismuth it is exceptionally high. This may be due to the very high nuclear magnetic field. This causes the Proton layer to tilt further away from the Neutrons diamagnetic field poles, if it reaches a full 90 degrees then the diamagnetic field would radiate outwards from the nucleus and be found setting at 90 degrees to an applied magnetic field always providing a repelling force external to the atom.

Also when we spin up a diamagnetic material like Copper and bring a magnet near it we see the diamagnetic field increasing with motion as well as a 90 degree torsion [tempic] force appearing. The 90 degree torsion force has been a mystery to me as such a motion is not explainable from only an electron magnetic interaction off the Copper cylinders curve. I would suggest that spinning the cylinder is causing the Proton and Neutron spin axis to align and forcing the magnetic fields to align as well, this produces a stronger repulsion between the Neutron and Proton layers and releases more of the tempic compression force of the Neutrons by tilting its component fields apart. Spinning a diamagnetic substance should cause the atoms to take on the Protons magnetic field as dominant. Now as we bring the magnet near the Copper atoms in motion the Protons field interacts pulling it away from axial rotation, and pushing the diamagnetic field [Neutron] the opposite way from the spin axis. The tempic field is 90 degrees to the voltage and magnetic vectors and in this interaction we finally identify how the three interact. As we decompress the magnetic field along its B vector in the Neutron we discover a torsion vector at 90 degrees to it popping out and interacting with the external magnets atoms to produce a negative motion force [dragging force]. This dragging force, and diamagnetic repulsion which seems to oppose any magnetic changes in the magnetic field, responsible for both induction and diamagnetism, is the Neutron.

Since the Neutron produces a tempic squared force due to the close overlap inside it, its tempic vector is much stronger although it still falls off as a linear function. Also its diamagnetic field would come out about 9 times stronger yet still fall off as a distance cubed function. If its voltage vector was not canceled it should be appearing at about 4 times the Proton and Electron level although still fall off at a distance squared rate. We see two of these forces appearing outside the atom, one is the diamagnetic repulsion and one is the dragging force [tempic field].

Strong Force

The strong force must be the result of the same 3 fields found above. It is tempic vectors of motion moving towards a parallel alignment and slowing the time flow rate at the center of the atom so that all motion along the tempic gradient which now increases outwards causes particles in motion to move slower along the inner side. Protons are not held in by voltage but by curved time flow rate or a spherical tempic gradient moving slower towards the center.

Wilbert Smith also indicated that when two particles share more reality or overlap [over half] the force interaction becomes squared. This means that the tempic interaction in a Neutron is far stronger then between the Electron and Proton layers and the strong force is now seen as an altered tempic vector squared function. Exactly how it gets to 137 times stronger is not clear but must have to do with the distances between layers.

The strong force is not Electric, it involves the diamagnetic field for repulsion and the tempic vectors for attraction.

Relating Time Flow Rate to Atomic Layers

Both the Neutron and Proton have the highest mass of the atom, one is seen trapping and compressing all three tempic lines of motion and the other is seen trapping or shrinking two. This does not explain how gravity is propagated between two atoms such that it may move around a magnetic field as well as somehow be linked to a lower time flow rate found mainly at the Nucleus. But it does pinpoint the source.

In our spinning copper cylinder experiments we saw a mysterious second force become present, the dragging force setting at 90 degrees to the diamagnetic field.

It produces motion [tempic vector interaction] and yet seems to travel across a diamagnetic field defying the natural laws of leverage and mechanical force.

Between the spinning Copper cylinder and the magnet held near it this mystery force is pushing off the diamagnetic field at 90 degrees inside the magnet and trying to drag the magnet along with the spinning Copper cylinder from an inch or two away as though it were held by a spoke from a wheel. As the Copper cylinder falls away on both sides there is no physical way to explain this with a magnetic field pushing between the two objects. If we were to build a physical lever system to create this effect we would need a fulcrum point setting in mid air between the two physical objects. So here we see that within the diamagnetic field from the Neutron is another tempic force resulting in motion moving across the field achieving a physical push at 90 degrees to its repulsive operation. The Neutron is

The Rain Maker Device

radiating a tempic or slower time flow rate which is thrown into the field on one side of the magnet and causing motion. The diamagnetic field is able to radiate a force flow differential in not one but two lines of space simultaneously at 90 degrees.

It is this tempic release of the Neutrons compression along two vectors of space that is being directed outwards through the diamagnetic field.

The Proton gives the Neutron one vector of release and the Electron gives it two. The magnet is repelled outwards and it is also dragged along and this is two physical vectors of motion simultaneously which can now be used to represent magnetism and torsion or time flow rate coupling.

Torsion is coupled through the diamagnetic field expanding outwards as the magnetic field moves inwards.

Reversing this torsion vector, to cause the drag to become thrust would require inverting the Proton and Neutron fields. That is forcing the Neutron to flip inside the Proton shell or at least to move past a 90 degree position or forcing the Proton shell to tilt beyond a 90 degree position while holding the Neutron vertical.

Crossing the 90 degree threshold would start to create an expansion of the tempic field outwards and a faster time flow rate curving gravity outwards.

How does motion through the Neutron Proton layers operate?

In the creation of a Neutron from an Electron and a Proton we see some changes to the radiant forces of each.

Both Electron and Proton are altered in size as the Electric vector pulls them together at the closest range possible. Proton grows larger and a conversion of mass to stronger magnetic field would be expected pulling it towards the Electrons properties a little. Electron shrinks and probably does the opposite. It gains a little mass and lowers its magnetic field. As the magnetic field increases with a distance cubed rate, and the electric field at a distance squared rate, the tempic field linearly, at this close range the magnetic field becomes nearly equal the electric field and it matches the electric field to pull the Proton to the center suspending the Proton at its center. The strong tempic vector spin couples and holds it in the center of the magnetic field as that is where motion is fastest. We now have two magnetic fields setting in the Neutron that are both somewhere between the Proton and Electron fields in attraction. The resulting particle does not radiate an Electric field, it radiates a small magnetic field, and a dragging torsion [tempic] field.

The balance of the two fluxes from each magnetic field creates the diamagnetic field.

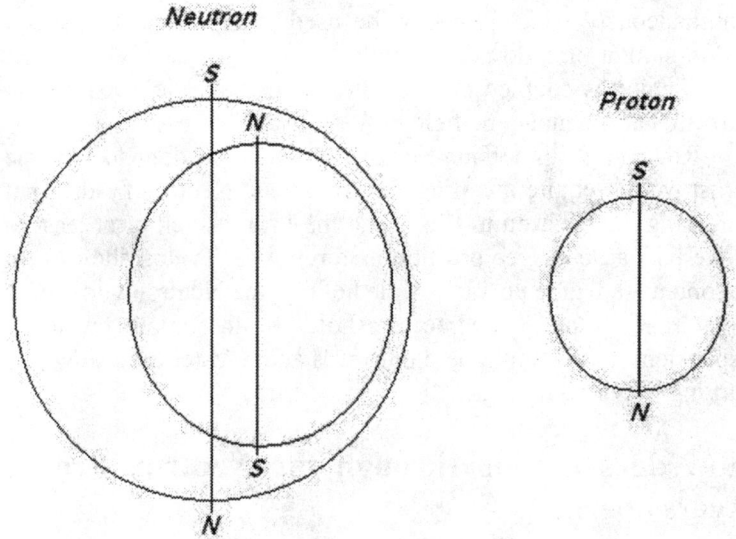

As a normal Proton comes near the Neutron it would normally be spin coupled by the torsion field and flip to oppose the outer Electrons magnetic field because it sits outside. This position would put it however in attraction to the inner Neutrons enlarged Proton which now has a stronger magnetic field, and the Electron has a lower magnetic field so the distance is much closer then the Electron shell of the atom. The flux diverted in the Proton interaction from the Neutrons outer shrunken Electron draws some force away from its inner flow to the inner Proton and now it's stronger field becomes involved in the interaction. Since magnetic fields both push and pull from top and bottom the magnetic forces cause a longitudinal shift inside the Neutron rather then a tilt. We could expect that each time the Proton orbits, a longitudinal wave would be created through the Neutrons density field radiating as a spiral.

Then the next layer the Electron orbitals [not pictured] will create a similar effect, only this one can be reversed with an NMR flip so will tend to cancel as NMR produces about 51% of either config at room temperature. This may explain some about superconductors forming at very low temps where all the Electrons would probably be flipped the same way. Regardless, this longitudinal force would seem to be Proton Neutron mainly. If we take the Protons Orbital velocity, circumference, and the longitudinal waves propagation velocity, we can determine the dimensions of the wave spiral and design a diamagnetic antenna to harness the longitudinal wave.

As to torsion interaction with the Neutron, we know that the Electrons orbital and particle spins are opposite, so that either way Electron flips it will still be effected the same as to tempic vectors. Proton is however different and if we could alter the Protons orbital with respect to the Neutron this would cause some serious tempic interactions because Protons orbital and particle spins are the same direction.

If we could flip the Proton while holding the Neutron we would get both a tempic expansion and also a magnetic repulsion between the two heaviest particles.

This would seem to be the greatest gain and would lower the diamagnetic field if not totally cancel it.

Searl Disc Revisited

It would seem all roads lead back to the Searl disc. The maximum tempic effect we could achieve would be to tilt the Proton magnetic layer away from the Neutron diamagnetic field or the inverse.

We know when we spin up a copper cylinder the diamagnetic field increases. This is probably from the Protons being pushed into more perfect alignment with the Neutron and dividing its internal magnetic flows.

We also know the Protons normally align with the Electron layer in one of two alignments, so they appear to be more coupled to the electron layer which has a stronger magnetic field then the Neutron.

It would appear that possibly it is the Electron and Proton magnetic fields that are staying vertical in the Searl disc and it is the Diamagnetic field that is turning to 90 degrees inside it which may be the actual operation of the copper atoms in the cylinder.

Spinning up the disc will artificially increase the size of the diamagnetic field which will seek to repel all magnetic fields outwards. Polarizing the Electron and Proton fields vertical will inhibit them from turning using a weak magnetic field. The 90 degree pulses will move straight through the electron fields with little effect, but when they hit the increased diamagnetic field it will be turned to oppose the 90 degree stronger pulsing field, and here we see the tempic lines of force moving away from a parallel path between Proton and Neutron. This would increase the time flow rate at the smallest awareness level of matter and begin to shrink the control fabric as it expands the atoms.

Lastly it is noted that the Protons precession frequency is lower then the Neutrons precession frequency so we can expect the diamagnetic field is able to turn much quicker and the Protons will lag the turn if they are caught in the pulse.

This model explains why the disc can be turned up either direction and loose weight in either direction of spin.

The difference in spin direction may reflect merely the temperature of the Copper and its NMR ratio of aligned or opposing flips to the induced vertical magnetic field.

Gravity

Wilbert Smith had identified the force of gravity as relating to orbital precession because he recognized it must be a distance squared force or voltage with a tempic modulation. He certainly never had access to the wealth of information that has led to a paper like this one identifying so many such atomic variables and assigning observed forces to particles. My best guess at this point is that gravity is the result of a propagation of the longitudinal

The Rain Maker Device

wave produced in the Neutron as the Proton orbits it. Thus in a Neutron star we would see Neutrons spin couple to one another and begin to orbit one another producing the strongest Gravity we have found appearing in our universe at present. One possibility I am not aware has ever been suggested, is that a Neutron star may in fact be very diamagnetic.

As the Neutrons center Proton is pulled off center by the Proton shells orbits what we see happening is an "Electric" effect generated along the spherical surface of the Neutrons inner Electron shell containing a tempic field of motion. The Positive charge balance to the Electrons negative charge balance is alternately compressed moving around the shell and a longitudinal voltage wave is thus transmitted out in a spiral. As the atoms mass increases with addition of Neutrons and Protons the waveforms would become faster and more spirals would emerge. This spiral wave would be radiated from all matter and thus when it intersects other atoms having Neutrons, those Neutrons would receive this wave and somehow transform it back into voltages and tempic force along the interfacing edge. That is the side the wave hits would become charged ever so slightly and contain a spin vector causing the atom to spin off centered and produce motion of attraction acting to accelerate the atom towards the longitudinal wave. This is a guess. The tempic vector and the voltage vector would both be present in this wave but the diamagnetic force would not.

Wilbert also suggested a method of skewing gravity outwards. Simply reverse the spin precession so that the tempic vector was reversed of the normal voltage vector. In this case we would simply have to rotate the Proton orbital spin onto an angle further then 90 degrees tilt to begin to reverse this longitudinal waves phase with respect to the tempic and voltage vectors. When the two waves emitting from two atoms intersected the tempic vectors would now be in opposition and repulsion would be the expected effect, based on the pattern from the conscious fabric. This also is a best guess and as longitudinal waves have not yet been even clocked or measured since Tesla we shall have to wait and see. It has been recognized early on that gravity at its source is coupled to the tempic field and time flow rate, however its propagation medium is still not certain. If this model is correct then gravity may be propagating along the Neutrons density and represent the mysterious scalar wave or Tesla wave, and this would explain its over unity flow medium. The tempic field rooted in the conscious fabric, the source of over unity and perpetual motion, and why Gravity never runs down or becomes depleted as a force always present in matter.

Just as the diamagnetic force of the Neutron is inexhaustible, and the Electrons magnetic field is never ending based on its tempic motion of perpetual spin, Gravity rooted in Voltage which is always a fixed quantum of electron volts and the tempic motion of the Protons orbits would be infinite but may be controllable as to spin flip orientation.

Diamagnetic Fields Without Generators

From Dave to Kosol :

Kosol,

No, we do not need the generators.

How hot can you make your hands become from consciously focusing?

If you make a bismuth coil with the weave winding on it and simply short the winding at the ends of the wires. Then place it into an iron pipe. Place strong magnets all the way around so one pole points into it.

The "diamagnetic" field will expand out and you can do the rest with you own powers of mind. It takes a lot of pure focus directing the mind into the bismuth coil, then things will expand out a whole bunch more.

The function generator makes it impossible to ignore and easy to do.

I have gotten a field to pop out using no function generators at all, and last night I did it again.

Here is how I did it last night.

Make a bismuth coil using the weave winding. Lay it sideways inside an iron pipe with a large enough radius for the coil to lay sideways. Put a normal coil around the pipe and connect the winding to the bismuth coil. Now place the magnets on the iron pipe all pointing in. Set a crystal sphere on top.

The energy is much easier to build using the mind and conscious link the better the device is made. Remember this stuff is really powered by the mind anyway. It just takes longer to build a field if a person is not strong with the diamagnetic field in their hands already or good at the mind link to the device. The function generator allows the mind to slide in faster and easier.

In order to get a higher time flow rate I believe we have to draw the diamagnetic field into "expansion" first because it is the deepest one in matter. Setting all the forces into opposition is the key. If you are good with vibrations already I believe this will work with only your hands to power it. But it will be harder to "learn".

Have you done anything with bismuth?

If I find a way to make it happen easier using my Bismuth I will be able to build some more of these and send you one to try. I am making a couple more of these today of different sizes to see which are more powerful and what wire sizes are best.

I have also discovered that the heat of the hands is not the only thing that these coils will respond to but all conscious intentions. "Divided intention leads to a slower time flow rate"

Every time we perceive conflict of intention so as to make a turn or a decision we slow our own progress. Perceiving all conflict, resolving the truth that lies in both opposing paths and absolving the conflict is the fastest way to speed the time flow rate. On the physical it is reversed, to speed the time flow rate we must set the forces against one another in balance. This is why the ring of magnets all pointing inwards causes the bismuths diamagnetic field to expand, and this speeds the time flow rate.

The Rain Maker Device

Is the device I see beside you in the picture Aluminum?

Very cool!

I know this is only a beginning that we are touching and it requires that mankind slowly move into a different perception layer on the conscious fabric to become more aware of how things really operate here in the physical world.

I do not know how the Joe cell works, I have no idea how a gas engine can run on water, but however Joe down in Australia is doing it, very few others can, if any. The difference is in the conscious fabric, not in the physical fabric.

The conscious technology does not hold Power for those in conflict of intentions. Built into nature itself is the protection. To harness the conscious power requires absolving conflict of intention. Thus this is not going to be used to power conflict between humans because in their hands it goes dead.

So to fire up one of these devices without using any function generator, expand you consciousness to the whole physical universe as one, and then focus into the device and vibrate it. The field pops out and then others can feel it too. It is just like a crystal only about 100 times stronger.

Dave Lowrance

David Lowrance Testimony

I hope you do not feel you are being ignored. I have read your posts and tried to respond but timed out so this time I am using my word processor.

"I still want to know what peoples experience has been, and most importantly if others are connecting with the same guardians for this technology as yourself, Kosol."

My comments:

I take this all with a grain of salt at present, as there is no way to "prove" it is real. However I feel it is accurate as I experienced it so I present it as such. Also others were involved at the same times and this is hard to ignore particularly in Kosol's case. At the risk of sounding crazy here goes.

Some of my perceptions and a little history:

For some years since I practiced Reiki Level Two, I have learned to connect with others at a distance. This is common teaching in Reiki but very few continue to use it at any great distance.
 Mary Ann turner gave me my first attunement from North Carolina, and I must admit it was as if she was present in the room. This was my first introduction.
 Some years later, as MaryAnn had suggested, I realized that as we come together on a site like this one we form a common energy point on the Mental and Astral planes of awareness. I spent a few years working with an emotional recovery site and learned it is very true. We begin to sense when someone is in need and are drawn to the board. The board forms a common connection point where we all spend time focusing and reading

The Rain Maker Device

one another's comment and thoughts. We are all corded in, with either astral or mental cords, or both. These planes of awareness are more real then the physical plane and we are building structure on them. Those who take this seriously create structure and those who scoff and peck away tear down structure, it is our choice.

The first time I started "sliding" with the crystal devices I nearly got lost, very disoriented, and finally my guidance brought me back and showed me the connecting point in the mental plane that 5 people on a Yahoo site had created. I immediately recognized it and I used it to get back here mentally and stabilize back in my physical senses. Someone was present thank God at just the right time to help me ground again, and the others also helped to get my physical mind working again. Well I must admit, now I cheat! I use the device to travel outward and connect with others. I do not know if this may be visible because of the new intensity of the device. It is Astral and Mental traveling.

I have not sensed too many others out there doing this, but last night or the night before I found one. The Spirit form was existing vertically to a great height above me as well as all the way down to Source. He became aware of my presence as I passed as well.

The first trip under:

I had invited my guides and the aliens to participate in my experiments for guidance.

Some time ago near the beginning, when I was out and perceived the aliens, they communicated with me telepathically. I received one message the first time, "Comprehension" was necessary for man to take the next step, thus the magnetism site and the data base may be some kind of a beginning. The second strong communication, when I had tried to do a physical transport using their device was "Do this between two earth points first". Do NOT be jumping into our machines. LOL! And indeed even Kosol came back to me and said I had scared them with that attempt because they realized I had learned how it works.

So I have been waiting for others to catch up with the scalar sliding to see if we can accomplish this goal.

The images are transmitted and received using the scalar receptors in the pineal gland that sits between the two hemispheres of the brain. Just like the Yogi masters have indicated. The scalar transmissions are very real, as though you are overlapping space in two places and both become visible. The aliens appeared in my room doing this, and at first I thought they were just a vision. I lay down to sleep, when I shut off the lights they became totally visible, because now the room was totally dark. As I layed down there presence did not tilt but stayed level with the room.

When I asked them to disconnect they did immediately. They had only connected because I invited them to previously at Kosol's suggestion. I figured what the heck, I can use all the help I can get right. I sensed them from time to time as I worked on the first paper, but it wasn't until I became aware of the device they were using to connect with me, and that experience in the bedroom, that I began to take them more seriously. It was a dodeccahedron with glowing points and it seemed as if it was setting off my left face about 6 inches away. I now recognize they had connected to me using an astral connection, because that was where I could work at the time.

Anyway I sensed they were showing me where we were heading, or could get to someday if we have time, and our civilization continues to use mind instead of just raw emotion to advance.

I have not connected with any aliens for some time now, and seem to have no desire at present, but have been absorbed in the information that seemed to be streaming through my mind on those first few intense trips. I think they are waiting on us to see if we will figure it out. It's up to us to comprehend and then prove we can do it ourselves first. This was the process started when Wilbert Smith was given the first scalar coils, that we have boldly embraced once again. I feel they have contacted me now as well.

My goal has been to help mankind realize we are all one conscious mass of souls connected together in an Astral Mental "cloud" and we must all move together or not at all. We cannot

The Rain Maker Device

enter divided or leave half ourselves behind and proceed into the "Mental birth." Enough people must reach "comprehension" to pull the entire cloud.

So in conclusion, whether I am delusional or truly inspired by a Spiritual perceptions that we are all heading for, is up to you to decide. I am not truly sure myself. LOL! But a therapist has told me the fact that I can question it shows I'm sane. I make no claims, and will truly be surprised if aliens start popping in after we succeed at scalar communications.

If aliens are real and not delusion, they are more spiritual then most of us. The aliens know that if we master this type of space overlap communication then we become aware of them simply by reaching out as I did and opening to them mentally. I think this is truly where the entry to the other races will have to begin. It is not "warp power" as on Star Trek. It is being able to see them, because the perception and control fabrics are more primal forces then the physical plane.

I still know that progress is slow, and every inch of ground must be understood by many people before it will become mainstream. Most of the serious people I have met on the sites was because of Kosol's continued presence and urging.

I set up the Kosol core tech site to be the center of experimenters that desire to experience the conscious energy first hand. I believe Kosol is a very sincere person with only one goal. So much of what he has seen appears to be confirmed by my own visions and perceptions.

"How we treat the least of us is as important as how we treat the best of us." "From the town drunk to the mayor we are all beautiful radiant Spirits participating in this experience known as life."

"It's all alive!" This is the mental plane perception of intense photons.

Ok there it is, my spiritual side of all this!

"In coming up with this God Device can we reverse the effects of aging, all health problems (or does this device work better for specific disorders) some are easier to work with like cancers, paralysis, while others more chronic and long standing require different methods."

I will leave this to the healers that are dedicated to this. I figure if God wanted us to live forever he would have set it up that way, but Kosol is fond of stating "No Limits!" I have perceived that the "diamagnetic" field responds to "intention" and built a device based on Kosol's guidance that seems to produce a very strong energy field that can be directed by the healer. My first perception of this was Chi as I learned it in the martial arts and acupressure. Who can fully explain Karma and healing? I used to believe "I" could heal others. At some point I finally woke up and realized that they heal themselves when they open to changing the root cause on the subtle planes and open to the Light themselves. The best we can do is point the way and show how it works by holding them in the Light for a time. Using an expanded healing field from a machine interaction makes one realize this is real. Everyone who has waved a hand across mine has concluded something real is happening, and it responds to intention.

"Is it much in asking for a basic materials list for rain maker construction?"

Right now all the devices that have been constructed have been a little different and development is still in progress. I have given a good record of everything I have constructed, however the experiment record is quite lengthy. This may be a good project for us to do later as others begin to work with the units.

For Example,

"1. Bismuth slug (where is the best place to get this) can it be bought at a regular hardware store?"

Bismuth can be bought at an ammunition store and is used in "shot" reloading. Or it can be ordered from a chemical supplier. None is produced in the US, it is all imported. I just posted a supplier called United Nuclear in the links section. A $20 bottle will make a couple slugs.

"2. Neodynamic 3/4" (what gauge/how many circular magnets?)"

Take a close look at the latest Rain Maker in the photos section under "Vortex Generator". There are only four stacks of the 3/4 magnets about 20 magnets in all. They are 1/8" thick. The

The Rain Maker Device

next size is 5/8" in 6 stacks of 4 ea around the top. The small stacks are 1/2" by 1/8" thick and there are 4 stacks of 8. The rest are fluff added later to extend the envelope a little larger. Check out K&J Magnetics and see if you can identify them. 20 ea DC2, 24 ea DA2, 32 ea D82.

http://www.kjmagnet ics.com/pullforc e.asp

This is about all you can get on the Ferrite core because they all push away from one another.

"3. Wire, how thick, how long scalar wire, where do you get this? best source for it?"

A local telephone company CO throws away loads of 24 gauge twin lead jumper wire. Some are up to 30 feet long. Other wise you can buy it. Every coil I have used in Rain Maker is less then 30 feet, but others are doing higher density work and using magnet wire which takes more talent because it has an enamel coating for insulation that must be scrapped off at the ends for connections. The bismuth coils are a single wire, and the ferrite is a twin lead coil having two wires.

"What about the measuring device which is shown on the photos, for measuring constant inflow and outflow?"

We have not had any success at a measuring device and this is still theoretical based on Tesla and Wilbert Smiths works. Hopefully time will give us some small devices. We still have yet to prove the faster then light T waves are real in an experiment.

"Ok, ok, it may be back to reading and I may be missing the obvious to you but consider it that I am technically challenged, just as you perhaps are language challenged. point me in the right direction and I will follow."

I wish I could send each and every person a completed device to work with and experiment! I do not have the means as of yet. Still trying to get Kosol up and online and looking for that first contact, a two way communication between earth points.

A little more background:

One of the first 3SD devices constructed was too powerful and this is not a good experience to have happen. Fear can shut down a project and make one never desire to return to it. Start with fewer magnets until you can sense these perceptions, and then slowly work to higher power levels as you learn how to channel it better and what to expect. Bigger is not better, and more powerful is not either.

I recommend the magnet stacks rather then cylinder magnets. The reason is control. You can vary the intensity by removing or adding layers this way. They are a better tool for experiments and can be used in other projects as needed. Also this creates a layering effect. The magnets are wrapped with copper and nickel in very thin sheets as they come from the factory. I do not know if this alters the field or not as I have no large cylinder magnets to compare them to. Certainly you can go with bigger cylinders if there is a cost savings, but you may want to keep some extra 1/8" thick ones for experimenting with the power levels. Pennies can be interlaced in a stack of neos, or the plastic inserts can be left in, but not in a cylinder magnet. You are committed to one config with the cylinders and one power level.

If your magnets are too big you will not be able to get them all on the iron core, they will begin to jump off. I have been very happy with the magnet setup I listed for this particular ferrite core. This is after weeks of playing with it in all different configs. I even tried placing disc magnets inside the iron ferrite ring which gave a minimal gain. Order a few extra ones, they do chip and break at times if moved around a lot.

Crystals:

On the crystal balls, size matters. I love my 5" calcite ball shown in the pics. Set my left hand on it and it rests comfortably. Each crystal ball is a little different of course, but any quartz sphere will vibrate the field. The smaller ones are more intense because of higher frequency. My Rain Maker does not use the magnet wrap on the large white crystal, it just produces and vibrates the diamagnetic field. I really consider the two projects different in this sense.

The Rain Maker Device

The smaller ones I have are dodeccahedron wraps around 2" diameter and an almost 4" blue synthetic crystal ball, very cheap, and also works well. A 2.9" Venezuelan crystal that is already hot to the palms with no magnet wrap. Right now I have an icosahedron set up on the 4" one and love the higher freq centered near the throat chakra. This crystal is amazing without any coils or generators.

Power:

The Rain Maker setup uses magnets pointing all inwards. This seems to affect the physical plane. The platonic crystal wraps on the spheres effects the conscious layers more, emotional and mental interaction, meditation and vibrations with perception. Spiritual growth. The magnet poles run along the surface rather then inwards. There is a remarkable difference between these two magnet patterns.

I hope this info helps people decide which direction to proceed first. If you want to make some rain you do not need the platonic wrap on the crystal orbs. If you want to prepare to channel the energy and not worry about rain then start with one of those first. A dodeccahedron wrap will help you iron out your astral problems first. An icosahedron wrap will open the higher centers. I would guess if someone wants to start talking with their celestial guides an icosadodeccahedron as appears on the 5th density, about 10" to 1 foot diameter. Pretty big crystal ball! Lots of magnets, very complex pattern to construct.

If you are into the hemisync idea then the Don Mitchel wrap.

Kosol seems to be wanting some rain! LOL! I would caution go slow, start little, work up. Take your time and observe at each step until you have grasped and become comfortable with it.

My experiments were laid out in order as I went through each concept. If you are missing a concept unpredictable results may scare you at some point. Each device gets stronger and relies on a comprehension of the previous ones.

Dave Lowrance

Kosol,

That coil around the ferrite ring is two wires wrapped on together. You end up with 4 wires coming out, 2 coils. I used a black wire and a white wire for easy identification of each coil.

Now you can hook the 4 wires up two or three different ways to test the difference.

To do a psi connection with the iron atoms connect them up opposite directions to the function generator. In theory this will strengthen mental function and clarity and pull consciousness back towards Source. Just as the inner scalar coil pulls you towards the Astral plane. This gives the entire conscious loop in one device for this density.

This coil on the ferrite ring also allows you to add a second frequency generator and get a mix of two frequencies operating in the field.

This unit has been a really good way for me to build my sensitivity to the magnetic energy, and to learn to link to the scalar energy.

On the other sites:

Also I can fully understand why Bruce and the other sites would consider this type of experiment unacceptable or not valid. This is where I stood 2 years back as well. I could not find any OU possible in this position. I had a Spiritual life separated from my scientific life, never did the two merge, thus neither were whole. My world was divided into two realities that were not joined.

That is why we now have this site to explore both aspects and try to find the common ground. I try to limit my posts on Bruce's site to the technical stuff I run into, because the conventional technology is what I am trying to interface with. Ultimately we need to discover a way to bring the Source energy back out into the physical to do work. These guys on the other sites have a lot of understanding on that side of the energy flow. EvGray has done some excellent development in motors, and if

The Rain Maker Device

you want to understand the inner working of flowing electrons and RF Bruce is a very good source.

Lawrence Rayburn has created some really neat scalar antennas and coils. There are a lot of good people still learning that side of the tech and searching for the answers in the entropy world. Not really thinking about where does energy come from? Where is Source? But thinking how can we extract it from somewhere else?

If and when we succeed at something here Kosol, then others will want the answers and begin to open their minds a little more. This will take some time to get people feeling the magnetic currents like Leedskilin could and I have just started to sense.

The currents will travel along a steel tie wire pulled into a small copper tube. Romag study suggested this. When I built a coil with this construction I have noticed it seems to draw the energy from Rain Maker into it. A diamagnetic wire. I am puzzling over this concept right now.

One more thing I must say, I am elated to be able to sense and feel these energies that I used to believe were only possible from human meditations and Chi work, coming out of a device. Even if we do not succeed at OU from this it has been well worth the efforts.

How is your device coming Kosol?

Seems like building devices never happens fast enough!

Some info for other builders:

I have had one other experimenter at c_s_s_p complete a Rain Maker device now and have completed one experiment. I have also conducted another experiment myself leaving the field active in the inflow for much longer to see what would happen.

It is our "perception" that this device has caused rain to manifest each time an inflow configuration is activated. So far 4 per 4 tries. I do not present this as a "claim" only an "observation." However this is enough to conclude much more investigation and learning may be in our future working with the "magnetic flow" that seems to be present in the Rain Maker device, and the "conscious" nature of the human interaction with it.

It has also been experienced now by two people that the field remains working until it is torn back down by magnet reversal. I must stress if you set up a field of the inflow polarity do not just forget about it and put the device away. Stay with it for several days and learn to feel the energy so you can reverse it when it is time. It may take longer to reverse the inflow field then it does to reverse an outflow field. I perceive the inflow is stronger with the same magnet strength.

The other device was built slightly different and this is where ones personal connection must always come into the picture. It was placed on a large metal bowl turned upside down which causes the field to expand more at the base of the unit. Also there was no crystal used in this device. The builder wishes to remain unidentified at this time for obvious reasons.

The result was not a surprise to me at this point.

As to the magnets on the Rain Maker base I recommend getting at least four of the 3/4" sets and then some smaller ones to fully cover the ferrite shell, these are strong magnets. Use a compass to prove that the poles around the whole device are not reversing anywhere. One pole is turned inwards and compressed into the center and the other pole is released outwards in the ring.

The center pole can only expand up and down with vibrations and the outer pole can expand in the other two directions. This causes any vibrations in the device to move vertically inside the device and horizontally outside the device setting up a quadrature relationship between the poles.

Some of my current thinking and alternate views:

My thoughts on Karma:

The subject came up "What if this falls into the wrong hands."

This is where we get into the Spiritual aspect of this technology. Most all religions have taught us some form of Karma "what you send out comes back to you." I believe that

The Rain Maker Device

karma can reverse evil intentions turning them back on the sender. A technology based on intention would force us to begin to understand the Astral plane interactions very quickly. Something spiritual people have already discovered.

If our civilization is to make another 30 years, and the oil is all gone, it will probably go one of two directions. Either back into tibal systems, or into a civilized society run by "mind" rather then "emotion" as it is today. I am certainly pulling for the second and I believe the rest of the world will follow if they begin to realize that what we send out to others affects us. We all go together into the same future.

I have to trust my inner guidance on this, at 52 I may not be here long enough to see the end results. My kids will however. I believe that if everyone is given a chance they will choose to be prosperous and peaceful, there is no one that is all evil. Everyone is justified to themselves. These are "perceptions" and they can be altered if we choose to turn to the path of "mind" rather then the path of raw "emotion". We cannot leave anyone behind, or benefit from hurting others. If we end up with all the remaining oil at cost to the rest of the world we will have karma to pay.

I think the younger generation is ready for this new perception of the world where power and control is replaced with heart and mind, and no one is considered worthy of sub human treatment. Could be I'm a dreamer, but the other option is to keep doing it the same, and I would even venture a prediction or two from this choice.

Alien communications:

Whether our alien friends are setting up there in ships waiting to help or not, this is beyond my perception. The posts from our spiritual people claiming to have this "perception" is unprovable by me. However they do make one thing very clear. It is our "perception" that is holding us back in the conscious technology. It is our perception that needs to be altered. Scientists do not need to be seeking proof and dissecting photos, but need to be seeking to duplicate and experience for themselves. Including "perception" itself as a variable is the only solution. Perception can affect the outcome of an experiment. Something right now is inspiring many of the older generation to seriously seek this new ZPE and AG technology. At 52 I have seen the world change, I now think more about the future of mankind then I ever did at 20.

Politics:

Our politicians are dumping money into war over oil, rather then into energy research. Even our scientists are acting "emotionally" and seek control over the outcome of an experiment to fit the known entropy models. People see a live demonstration of an OU device and leave the building trying to figure out where the real generator is set up refusing to believe there could be something we do not yet understand and unwilling to entertain a rewrite of the entropy models. Daniel Pomerleau is a good example. These "perceptions" need to be altered.

The stock market rewards people for speculating successfully rather then actually producing something real. We are becoming a nation of speculators because it pays better.

Opened science:

A science that does not write and legislate "Laws" as though it were a field of law rather then of discovery. The ability to consider all events observed, rather then only events that can be controlled the same everywhere and fit the current models. The acceptance of a [Source] concept. All energy comes from somewhere we do not yet understand. Not everyone can perform an experiment and succeed at the same outcome where "diamagnetism" is involved. Intention is a variable of the experiment and this is OK if we recognize it. There is no such a thing as a closed system, all atoms are connected to source and therefore have conscious aspects and interactions.

Summary:

I have chosen to present my finding on Rain Maker as "perceptions" rather then claims or laws. This is a direct result of

Wilbert Smiths charting of the Perception and Control field fabrics which I have attempted to make more sense of in my latest model on Creation. I believe these fields are real from what I have experienced.

I believe this is the only sane approach. There are many other models out there today to try on and see if they work, but fear must be dropped as it is simply an Astral manifestation that distracts mind from discovering truth through observation and experience.

Science must accept the interaction of both emotion and mind to begin to duplicate what Spiritual adepts have already found possible. Right now levitation is an art, for it to become a science we need to accept all the aspects of those achieving it. Any five people can achieve it together for a time by joining there focus in what is commonly ignored by scientists "party levitation". This has been observed for centuries, and scientists walk away denying that there is a "conscious" interaction worthy of study. This is called a "perception blockage". A desire to control the end result of an experiment and choose to be blind to the facts that are unacceptable to the mass of sleeping people as established by "Laws" dictated and enforced by science.

Trust in Spirit,
Dave Lowrance

Final Thoughts

"Well everyone here is some guidance to use the Rain Maker into making water of life and creating healing water, all you have to do is put a glass filled with water and put a small crystal quartz orbs into the glass also, then set the glass on top of the crystal device or next to it, and the scalar and consciousness field will energize the water and spin it atom and molecule into a threshold that give life energy, so whoever drink this chi charged life force water will be heal and energized and also their brain will increase in activity and their body and mind will vitalized.

As well for those of you who wanted to modified the rain maker into making zero power source system and antigravity propulsion. all you have to do is get to hemisphere of two kind of metals for example it can be copper and aluminum, or brass and aluminum just any two kinds of metals, then you separate this hemisphere with dielectric like plastic glue stick insulators or barium tenate can be used also, then use this dome like hemisphere to cover the Rian Maker device, connect two node wire to each of the hemisphere, and you can connected the two wire to the circuit that is connect to your house.

Once the rain maker active you will have both free energy power source or zpe (zero point power source), and also you will have antigravity because of the biofield brown effect from the dielectric between the two different metals (aluminum and brass, or copper and zinc or aluminum and copper etc.) domed like hemisphere that cover the Rain Maker.

Also, you can have a sphere of multi layer with two different metal encasing the entire Rain Maker device instead of just covering it.

So you see no limit whatsoever, but again you have to connect the two wire node. One wire to each sphere or one wire to each hemisphere, remember if you doing the sphere version you have to also separate the hemisphere with dielectric also at the equators, that mean the upper hemisphere (north metals hemispheres) is separated from the south metals hemispheres.

Well any question, you can join me on my forum links.

http://tech.groups.yahoo.com/group/Kosol_Core_Tech/

Also you can press on the links to get the website for the supplies to build the rain maker device.

http://tech.groups.yahoo.com/group/Kosol_Core_Tech/links

As well you can get my other books that I wrote:

http://www.e-booktime.com/Merchant2/merchant.mvc?Screen=CTGY&Category_Code=MP

Also my book is in Amazon.com, BarnesandNoble.com, as well as many other online bookstores.

I wrote, Star Gate Ascension, Cultivate Inner Force and Reading People Like a Book, Kosol and Keoun Spherical Generator, the Rain Maker Device, and more books to come in the future.

Ok, I see you in the forum and please help buy my books so I can raise funding to build more projects and write more books and also build meditation and healing centers around the planet to help humanity to meditate and ascend.

So thanks for everyone's support and may God and The Divine mother/father God bless every one of you.

Thank you from Kosol Ouch, Koeun Noun Ouch, David Lowrance, Vince Penala, Daniel Nissen, Jake Tepac, Carlos Sanchez, Jessy Sanchez.

Kosol and Koeun Ouch, Vince Panella and David Lowrance

Thanks to my mother, Heng Vun Ouch, my father Cham Nam Ouch, my sister Nancy Ruby Ouch and thanks to everyone who supports me and the project of planetary ascension into the fifth and sixth plane of density and beyond.

God bless.

Photographs

The Rain Maker Device

The Rain Maker Device

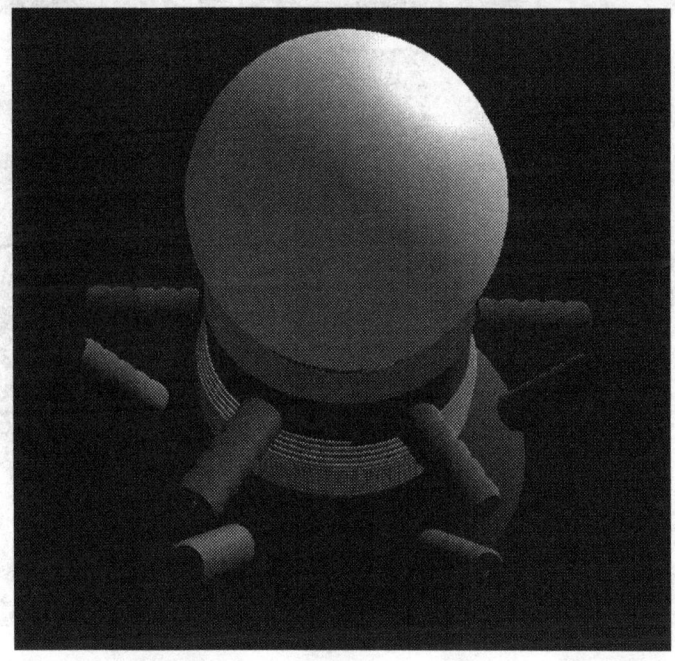

The Rain Maker Device

The Rain Maker Device

The Rain Maker Device

The Rain Maker Device

The Rain Maker Device

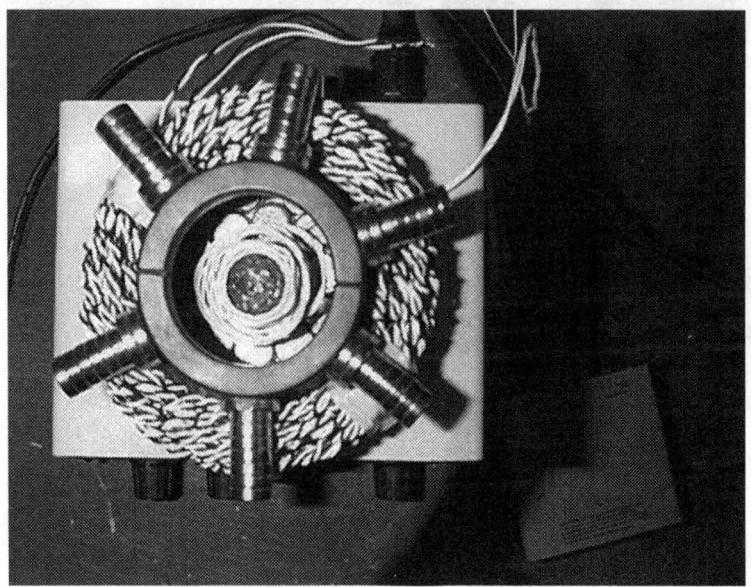

The Rain Maker Device

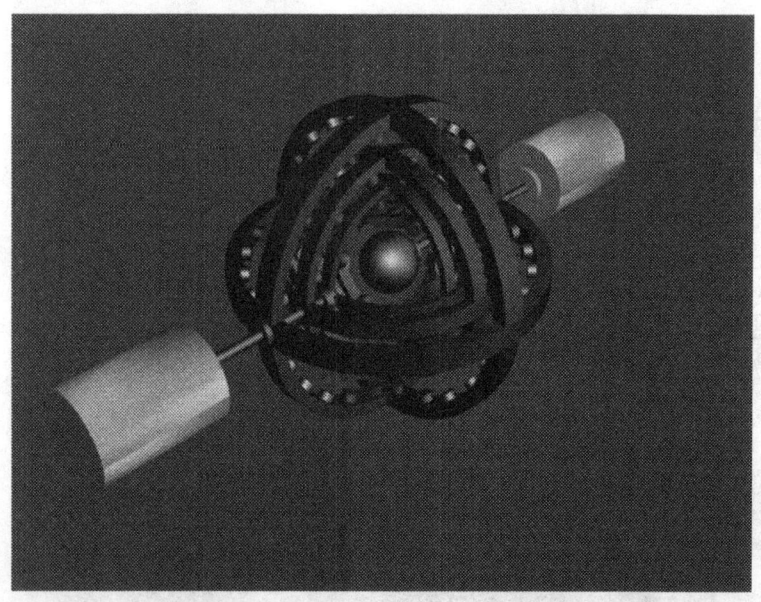

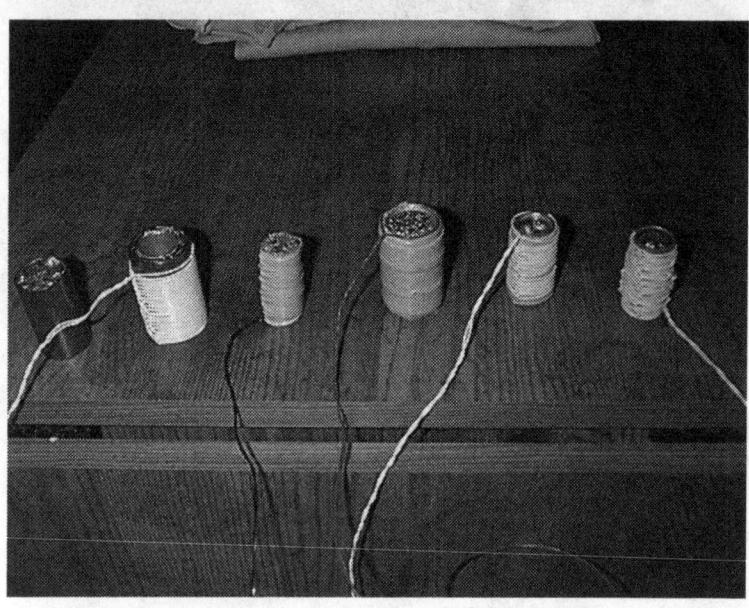

The Rain Maker Device

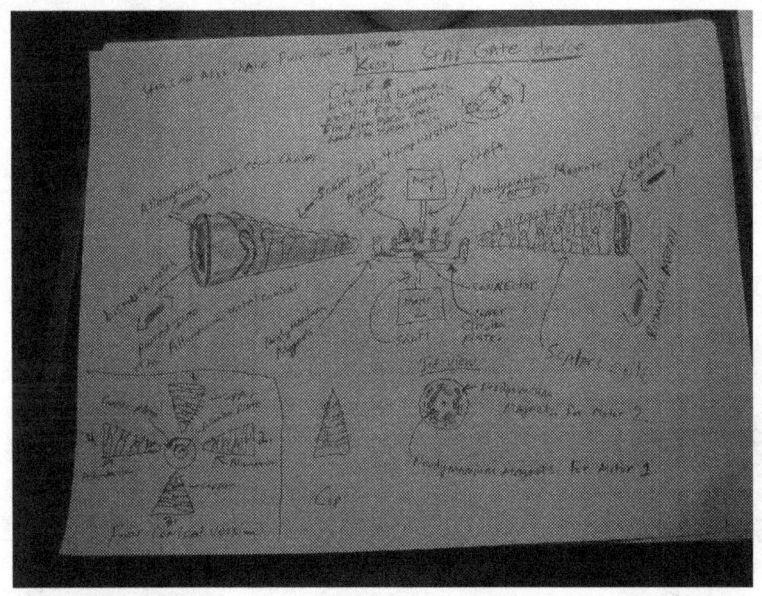

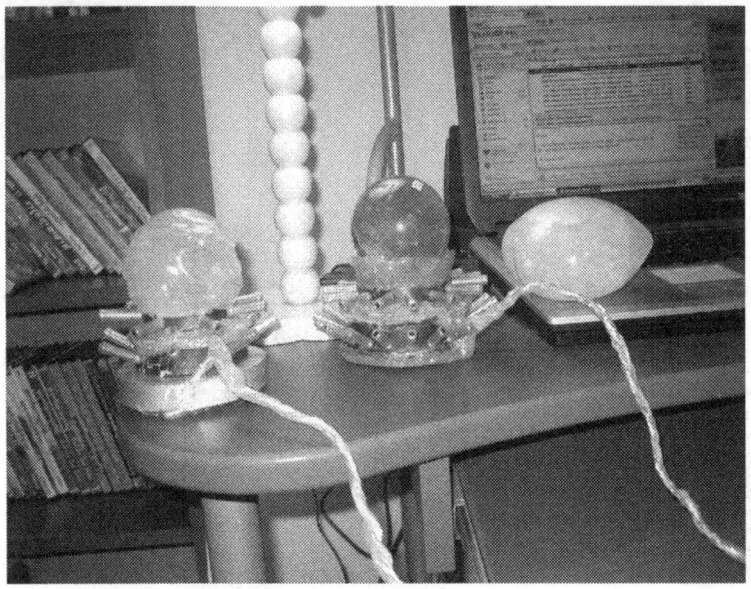

Kosol and Koeun Ouch, Vince Panella and David Lowrance

The Rain Maker Device

Kosol and Koeun Ouch, Vince Panella and David Lowrance

The Rain Maker Device